SHIPS, CLOCKS, AND STARS

Published under the title *Finding Longitude*
in the United Kingdom in 2014.

Ships, Clocks, and Stars

© HarperCollins Publishers 2014
Text © National Maritime Museum, Greenwich, London 2014
Authors Richard Dunn and Rebekah Higgitt
For photographic copyright information, see page 255.

In association with
Royal Museums Greenwich, the group name for the National Maritime Museum,
Royal Observatory Greenwich, Queen's House and Cutty Sark
Greenwich
London SE10 9NF
www.rmg.co.uk

All rights reserved. No part of this book may be used or reproduced in any manner whatsoever without written permission except in the case of brief quotations embodied in critical articles and reviews. For information address Harper Design, 195 Broadway, New York, NY 10007.

HarperCollins books may be purchased for educational, business, or sales promotional use. For information, please e-mail the Special Markets Department at SPsales@harpercollins.com.

Published in 2014 by:
Harper Design
An Imprint of HarperCollins*Publishers*
195 Broadway
New York, NY 10007
Tel: (212) 207-7000
Fax: (212) 207-7654
harperdesign@harpercollins.com
www.harpercollins.com

Distributed throughout the world by:
HarperCollins*Publishers*
195 Broadway
New York, NY 10007

Front cover images:
The Great and Newly Enlarged Sea Atlas, by Johannes van Keulen
(Amsterdam, 1682), PBD8037 © National Maritime Museum
Two English Ships Wrecked in a Storm on a Rocky Coast, by
William van de Velde the Younger, *c.* 1700, BHC0907 © National Maritime Museum
Endpapers image:
A map of London, from *A New General Atlas, Containing a Geographical and Historical Account of the World*, by John Senex
(London, 1721), PBB5531 © National Maritime Museum

ISBN 978-0-06-235356-6

Library of Congress Control Number: 2014940259

Printed in China

First Printing, 2014

SHIPS, CLOCKS, AND STARS

PRODUCED IN ASSOCIATION WITH THE *Ships, Clocks & Stars: The Quest for Longitude* EXHIBITION
AT THE NATIONAL MARITIME MUSEUM, PROUDLY SPONSORED BY UNITED TECHNOLOGIES CORPORATION.

Richard Dunn & Rebekah Higgitt

Contents

6–7	Forewords
8–11	Prologue: A WORLD DIVIDED
12–33	Chapter 1: THE PROBLEM
34–65	Chapter 2: THE CONTENDERS
66–103	Chapter 3: ON TRIAL
104–125	Chapter 4: MAKING LONGITUDE WORK
126–157	Chapter 5: WORKING AT SEA
158–187	Chapter 6: COMMERCE AND CREATIVITY
188–221	Chapter 7: DEFINING THE WORLD
222–225	Epilogue
226–243	References
244–246	Bibliography
247–254	Index
255	Acknowledgements and Picture credits

Ships, Clocks & Stars: The Quest for Longitude

Director's Foreword

Longitude is central to the stories of navigation and discovery told by the National Maritime Museum since it opened in 1937. It gained even greater importance to us in the 1950s, when we assumed responsibility for the Royal Observatory, itself founded expressly to help solve the 'longitude problem'. It is most appropriate and a great pleasure, therefore, to be able to commemorate the tercentenary of the first Longitude Act of 1714 with an ambitious exhibition and this new book, both of which tell what is an extraordinary story for twenty-first century audiences.

The famous Harrison timekeepers are, naturally, central to it and have been a draw for visitors since coming to the Museum in time for its opening in 1937. In thinking about them in this tercentenary year, it has been fascinating to look afresh at the often fraught events that first brought them to Greenwich nearly 250 years ago, and at their broader context as part of the longitude story as a whole.

I would like to thank the authors for their efforts in researching and writing this book. Their work has been possible thanks to a major research project on the history of the Board of Longitude, run in collaboration with the Department of History and Philosophy of Science at the University of Cambridge and funded by the Arts and Humanities Research Council, to whom we are most grateful. I should also like to thank United Technologies, which has supported the exhibition so generously. Together, the book and exhibition present an extraordinary story of innovation, creativity and competition that changed how we understand our world.

Dr Kevin Fewster am frsa
Royal Museums Greenwich

Sponsor Statement

Innovation is timeless. Yesterday's ideas form the foundation for today's inventions, which power tomorrow's solutions. At United Technologies, we are proud of our long history of pioneering innovation to make modern life possible. We understand the relentless drive of those who sought to solve the longitude problem. It's the same drive that pushes us to solve today's global challenges. In this spirit of innovation, we are delighted to sponsor the exhibition *Ships, Clocks & Stars: The Quest for Longitude*. We hope you are inspired by this great story.

Louis R. Chênevert
Chairman & CEO,
United Technologies Corporation

Ships, Clocks & Stars: The Quest for Longitude

Foreword

The Longitude Act of 1714 was an extraordinary event, an unprecedented moment when natural philosophers put a scientific problem on the political and national agenda. Their success was evident in the speed with which Parliament took up the call to action, and in the large rewards that the Act offered – sums that could be life-changing for the winners. More important, the potential rewards would incentivize energetic and ingenious efforts to meet the challenge, and the measurement of longitude was indeed the number-one technical challenge for a maritime nation.

The Act was also notable in creating a diverse group of experts, the Commissioners of Longitude, who brought together Britain's naval, political, academic and scientific interests. The Commissioners constituted what was in effect the first 'research council', aimed at rewarding invention and innovation. And although it is best known for the long-delayed recognition of Harrison's achievements, the Commission remained in existence for more than a century, rewarding other ingenious inventions and explorations.

The ex-officio members of the Commission included the Astronomer Royal, the President of the Royal Society, and a Cambridge professor. As the fifteenth Astronomer Royal, as well as a former holder of the other offices, I have special historic links with the Commission ('Astronomer Royal' is now, however, just an honorary title, without any formal link with Greenwich). I am therefore delighted that the 300th anniversary of the Longitude Act should be marked by a splendid exhibition at the National Maritime Museum. This fine book accompanies the exhibition. It tells the story of the search for practical ways of determining longitude while on a ship at sea, a quest that many considered to be as hopeless as the search for perpetual motion or eternal life. Yet the problem was effectively solved in the eighteenth century, largely by British artisans and philosophers.

This book takes a broad view of the subject, tracing the history from the attempts of the sixteenth and seventeenth centuries, some of which seemed genuinely promising, to the mid-nineteenth century, by which time new techniques for measuring longitude at sea had been embedded in naval routines. These advances helped create a better understanding of the world through improved charting, in which British surveyors and ships were a major force.

The story is also about problem-solving – the process of identifying a problem, exploring different options to overcome it, and then bringing workable solutions to a state where they can be used by all. Clock- and watchmakers including John Harrison, John Arnold and Thomas Earnshaw, and astronomers including Edmond Halley and Nevil Maskelyne, all feature prominently. But it is also a story that shows that the most difficult technical problems are not solved instantaneously: they usually require huge efforts over a long time to become a part of everyday life, often necessitating what we would now call 'public/private partnership' whereby the state offers support to inventors and entrepreneurs. Thanks to the priority given to the longitude challenge, London became a crucial centre for the development and discussion of ideas, instruments and techniques that would underpin major changes in seafaring, which was Britain's lifeblood.

MARTIN REES, ASTRONOMER ROYAL

A World Divided

it is well known by all that are acquainted with the Art of Navigation, That nothing is so much wanted and desired at Sea, as the Discovery of the Longitude, for the Safety and Quickness of Voyages, the Preservation of ships, and the Lives of Men.

'An Act for providing a Publick Reward for such Person or Persons as shall discover the Longitude at Sea' (1714)

In 1494, Spain and Portugal partitioned the world. Under the Treaty of Tordesillas, signed that year, a line 370 leagues west of the Cape Verde and Azores islands split the globe from pole to pole. Lands discovered to the west of the line would belong to Spain, those to the east to Portugal. East–west position – longitude – had become territorial. Yet the treaty did not explain which of the islands was to be used to determine the line's position, or how to translate leagues (roughly three miles) into degrees and so decide whether new discoveries lay to east or west. Portugal also assigned more leagues to a degree of longitude than did Spain, placing more territory under its domain. Moreover, the Treaty had effect only in the Atlantic hemisphere and things became even more difficult when both nations reached the East Indies. Within a few years, matters came to a head there over possession of the Moluccas, the 'Spice Islands'. The struggle for the control of the lucrative spice trade was intense, and the conflict between Spain and Portugal was only resolved in 1529 by the Treaty of Saragossa, which specified an equivalent dividing line in the East. Global positioning was, even then, a serious political matter.

This book is an account of how the determination of longitude at sea became feasible, and of how global positions could be agreed and the world known with greater clarity. On the one hand, it is a tale of seafaring, time and astronomy; on the other, it concerns commerce, competition and conflict, exploration and empire. The 'longitude problem', as it has become known, was a technical challenge that taxed the minds of many of the great thinkers of the Renaissance and Enlightenment. Galileo Galilei, Christiaan Huygens and Isaac Newton all grappled with it as a puzzle that seemed insoluble. Finding the longitude became a ridiculous quest only to be undertaken by the deluded, until the simultaneous development in the late eighteenth century of two practicable, complementary means of fixing a ship's position changed everything. These methods gradually came into use, both for routine navigation and for creating better charts of the world's oceans and coastlines, mapping the Earth in ways that had been inconceivable in 1494.

The quest for longitude is an international story, and this account touches on important work in the Netherlands, France and other countries from the late fifteenth century onwards. However, the main focus is on events in Britain from the early eighteenth century to the middle of the nineteenth. It was in Britain that the rewards offered under the Longitude Act of 1714, and the creation of an administrative structure to support promising ideas, led to the testing and development of the two methods that would eventually come into standard use at sea.

Why it should have been in Britain that the problem was solved is one of the issues this book addresses. The answer has much to do with the relationships operating between government, commerce and science at the time. Longitude solutions were encouraged by the British state through the 1714 Act, as had happened elsewhere; but, crucially, the new incentives addressed a British audience of skilled, commercially driven artisans working in a context of public discussion of new ideas. The Act therefore played to the strengths of Britain's metropolitan culture of craft skill and open intellectual debate.

Longitude mattered greatly at sea, but much of the story of how it was found and then deployed took place in cities on land, among academics and artisans. Crucially, the Commissioners of Longitude named in the 1714 Act eventually took on the role of encouraging promising work over many years, and of fostering the means by which the new techniques could be used on all ships, not just Britain's alone. It was not simply a matter of paying a reward; good ideas needed to be turned into reliable tools. Once they had been, Britain's existing maritime dominance allowed its navy to lead efforts to deploy the new methods for finding longitude in order to chart the world with certainty. As a result, a new line, now passing through the Royal Observatory at Greenwich, would come to divide the globe and define every ship's longitude.

(following page) – A map of the world, by Paolo Forlani, published by Fernando Bertelli, 1565

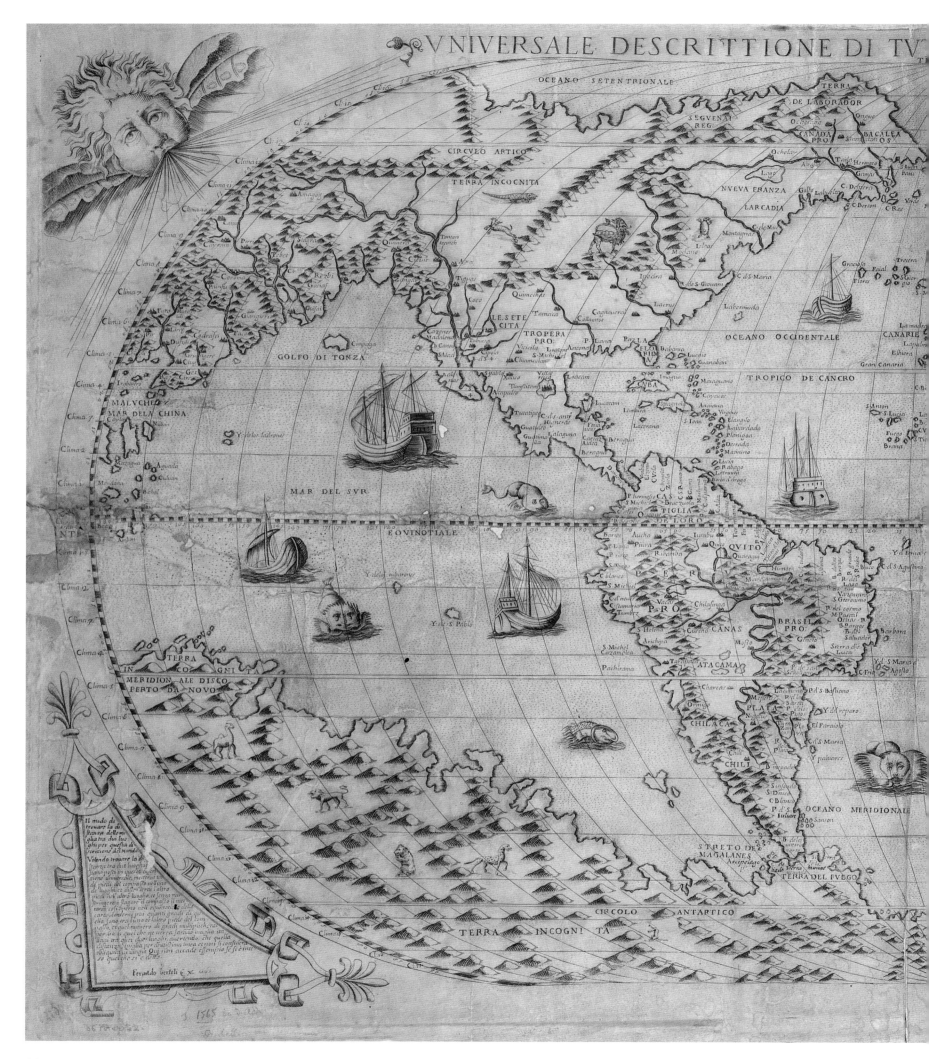

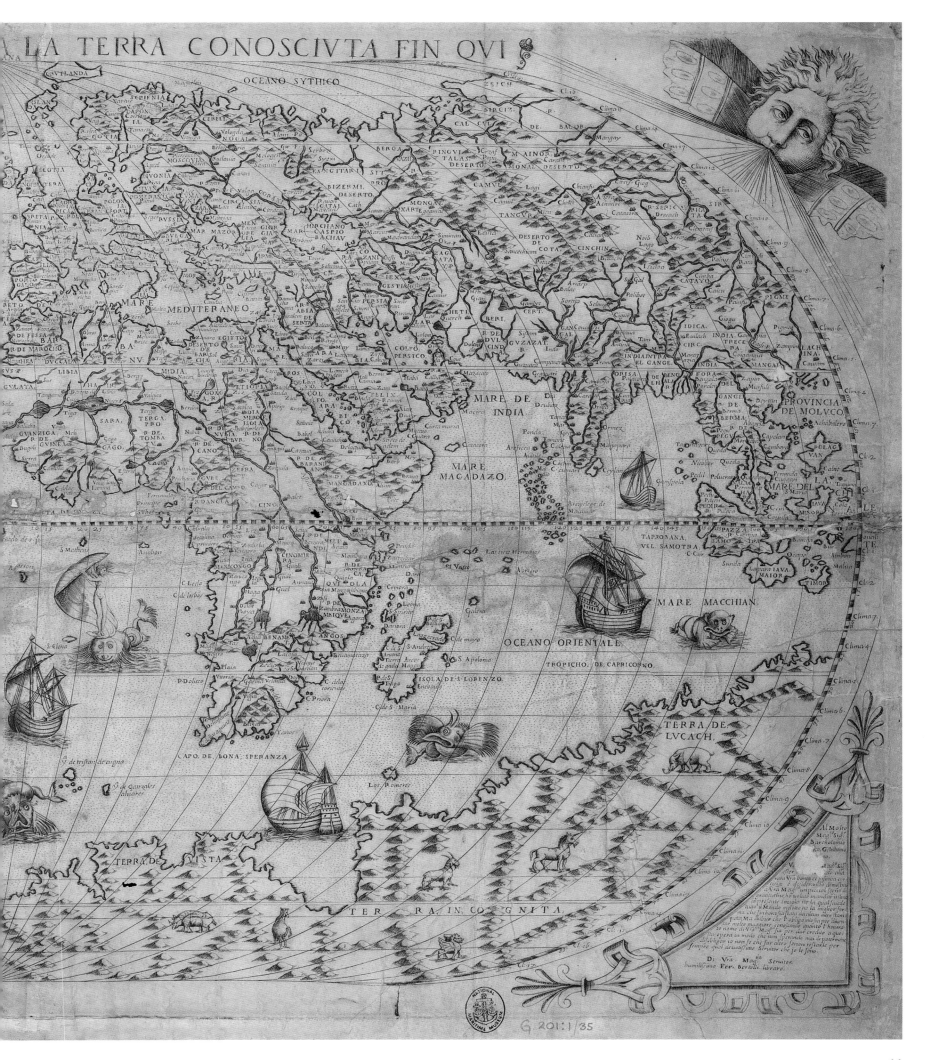

CHAPTER 1

THE PROBLEM

Nowe some there be that be very inquisitive to have a way to get the longitude, but that is to tedious.

William Bourne, *A Regiment for the Sea* (1574)[1]

Seafarers have always needed to know where they are to avoid danger and ensure a successful voyage. First and foremost, this was about safety, although they appreciated that more precise navigation could increase speed and efficiency. To most, this meant pinpointing the ship's latitude and longitude on a reliable chart. Latitude was fairly straightforward to measure from a ship. Longitude was the problem and good charts could only be produced when both could be measured.

As European vessels made longer and longer voyages from the fifteenth century onwards, navigation, including the determination of longitude, began to matter more. Long-distance trade, in particular, drove the desire for speed and reliability, and with it navigational certainty, to make voyages safer and more profitable. As international trading networks developed, and with them the need for stronger navies, navigational knowledge and training became more important to those with commercial and political power. Yet, despite this growing interest, the problem of determining longitude at sea would challenge seafarers, artisans and men of science for centuries before being solved, in principle at least, in the mid-eighteenth century. In the meantime, and, indeed, for long afterwards, seafarers relied on knowledge and techniques that had been developed over generations. Many voyages were successful, some ended in disaster.

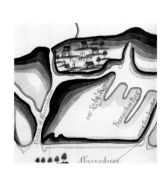

... some difference arose between them about Latitude and Longitude; Mr. Kempthorne alledging that there was no such word as Longitude; after that, further angry words arose

Evidence at the trial of John Glendon, convicted of the manslaughter of Rupert Kempthorne at the Ship Tavern in Temple Bar, London in October 1692[2]

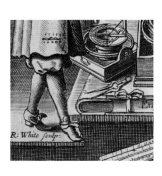

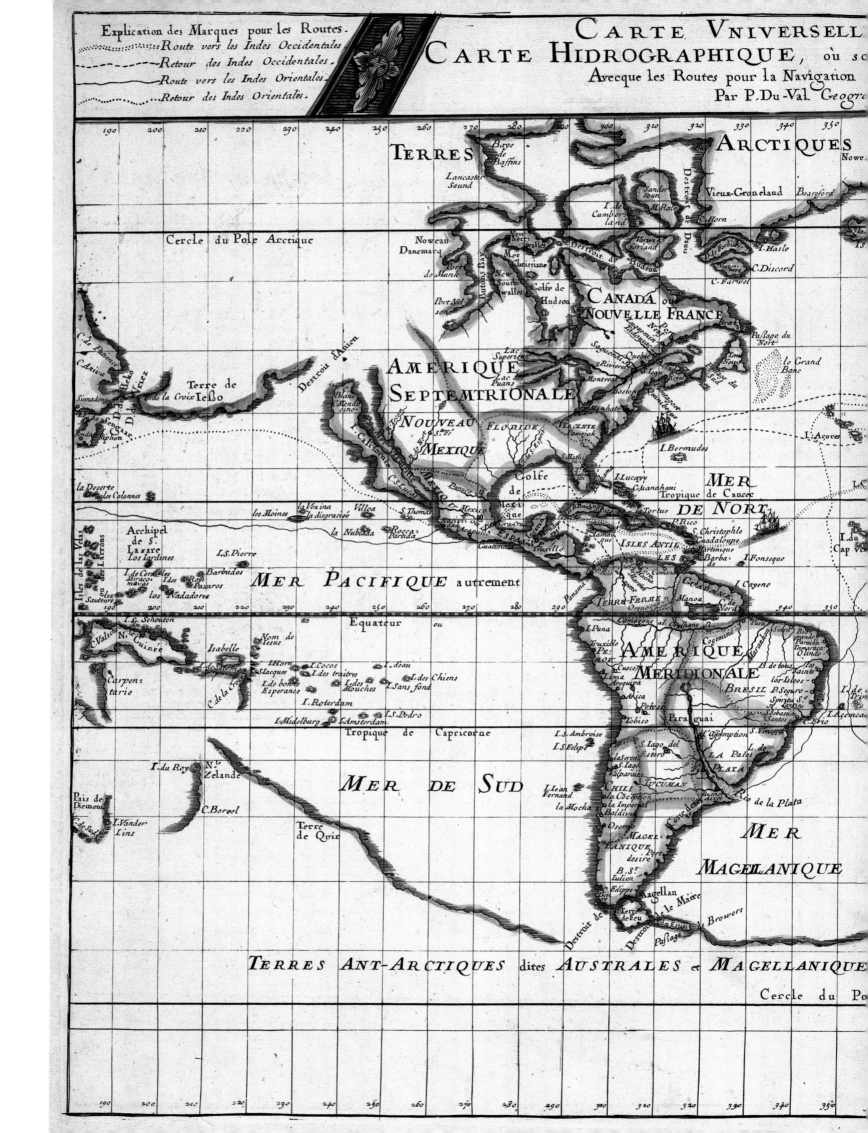

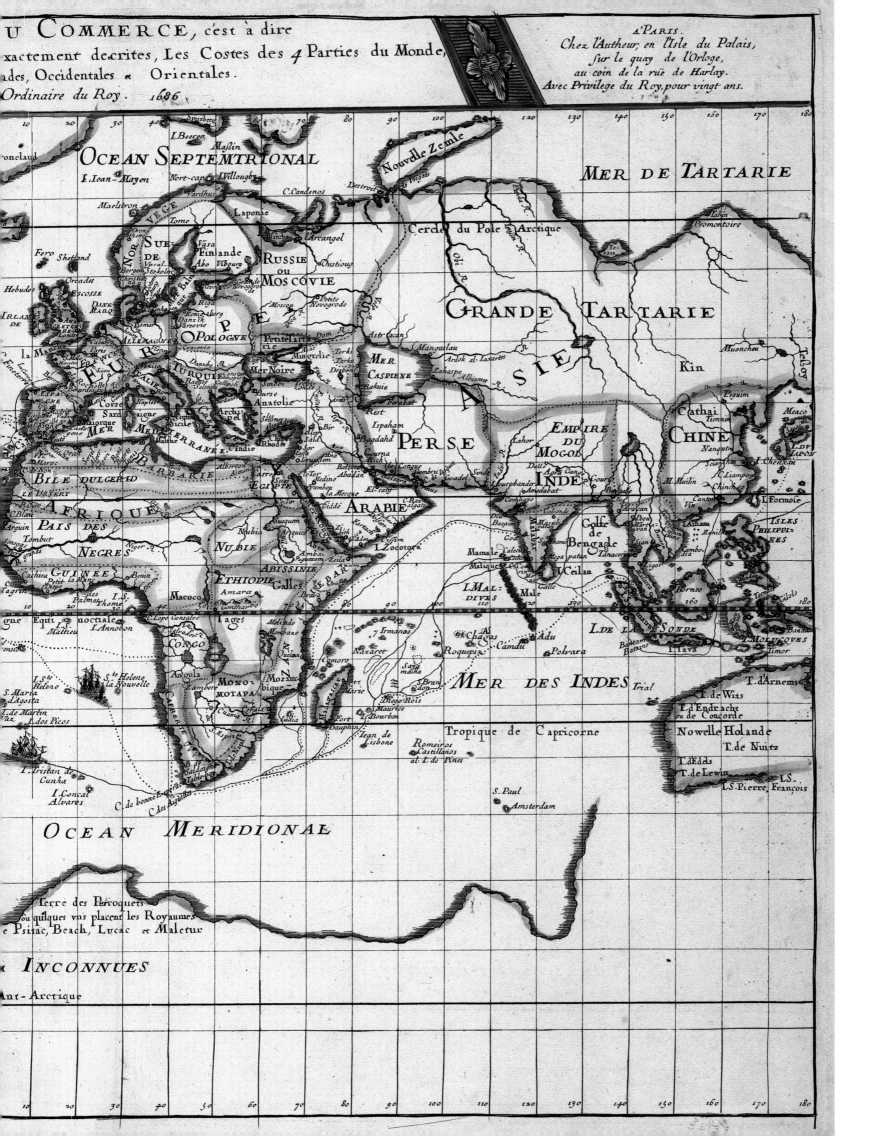

Why longitude mattered

The importance of being able to measure longitude at sea was inextricably linked with wider issues of marine navigation and safety. For many seafarers, the main concern was not simply a matter of getting from place to place, since by the seventeenth century it was possible to sail to many parts of the world with some confidence of return. Rather, it was whether this could be done more predictably, more quickly and with less risk; in other words, could it be done more profitably?

Broadly speaking, the further people wished to sail, the greater the risks, whether along well-travelled routes with known hazards or into relatively unknown waters. The determination of longitude and other potential advances were of most interest, therefore, to nations investing in long-distance trade and outposts and settlements overseas (Fig. 2). Having opened up trade routes to the Pacific and Indian oceans, Spain and Portugal were the maritime superpowers of the sixteenth century. By the end of the seventeenth, the Netherlands, France and England were coming to dominate the oceans. It is no coincidence that the chronology of rewards for longitude solutions mirrored this sequence of maritime activity.

The expansion of global trade was linked to a progressive rise in the numbers and activities of chartered companies. Britain's Muscovy Company (chartered in 1555), East India Company (1600), Royal Africa Company (1660) and Hudson's Bay Company (1670) competed with similar institutions from other European countries,

Fig. 1 (previous page) – *Carte universelle du commerce*, by Pierre Du-Val, Paris, 1686, showing French and Spanish trade routes to the West and East Indies. Note that longitudes are shown from a meridian through the Canary Islands

notably the Vereenigde Oost-Indische Compagnie or VOC (Dutch East India Company, 1602) and the Compagnie Française pour le Commerce des Indes Orientales (French East India Company, 1664). Subject to state supervision, each was granted the right to colonize, sign treaties, make and enforce laws, and hold a trade monopoly for specific territories overseas. The companies were largely free to do as they pleased but could draw on naval support and possibly, in times of crisis, government aid.

This was big business. In 1636–37, an inspection of the Spanish Manila galleons heading from the Spanish East Indies (Philippines) to New Spain (Mexico) valued their cargo at one million pesos (equivalent to £200,000 at the time and over £17 million today), while, in 1685, a French observer claimed that Dutch and English trade with Asia was making profits of between twelve and fifteen million livres (around £10,000,000, or more than £870 million today). This was exaggerated but English imports of tea, coffee, spices, textiles, chinaware and other commodities from Asia have been valued at just under £600,000 for that year, while the loss of five East India Company ships to privateers in 1695 cost the company £1,500,000. (Privateers were privately owned ships that had state permission to attack ships of enemy countries – and to keep the plunder.)

Privateers were just one of the risks. A ship's high-value cargo was also in danger from natural hazards, such as storms, throughout a voyage, as were the lives of its crew. Between 1550 and 1650, one in five ships was lost between Portugal and India,

Fig. 2 – A busy Dutch East Indies factory port, possibly Surat, by Ludolf Backhuysen, 1670. Dutch and English ships can be identified by their flags, testament to the commercial interest that both countries had in Asia

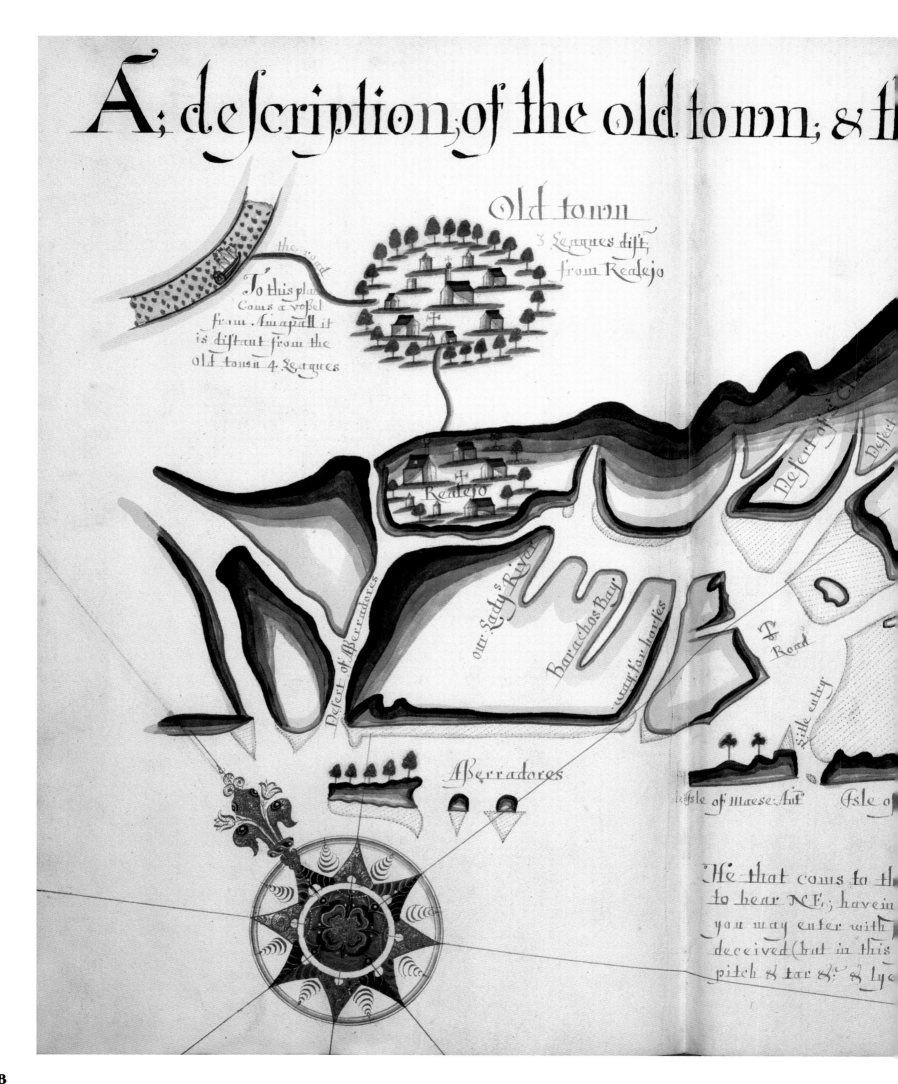

and crews had a one in ten chance of dying during the voyages. It is no surprise that the safe arrival of a trading vessel at remote outposts was a cause for celebration, or that sailors looked to protective measures such as amulets to keep them from harm.

Each of the main trade routes – between Europe and America across the Atlantic; between Europe and Asia around the Cape of Good Hope; and between the Philippines and Mexico across the Pacific – presented its own challenges. Stormy passages in the Strait of Madagascar plagued Portuguese and Dutch vessels between Europe and Asia. The Dutch established an alternative route in the seventeenth century, sailing eastwards from the Cape of Good Hope until reaching the correct longitude and then turning north towards the trading posts of Indonesia. If they sailed too far east, however, they were likely to fall foul of the reefs of Australia's western coast. It was a route on which knowing longitude really mattered.

Trading companies and the navies that supported them clearly had a vested interest in better charts and improved understanding of sea routes. As the famous diarist and naval administrator Samuel Pepys noted in 1683 in his Tangier Papers:

> the East Indies masters are the most knowing men in their navigations, as being from the consideration of their rich cargoes, and the length of their sailing, more careful than others ...[3]

The companies encouraged their officers to gather data about weather patterns, currents, coastlines and sailing directions. It could be sensitive information. In the sixteenth century the Spanish monarchy prohibited the circulation of maps and descriptions of the Indies to protect their outposts in the Pacific. So it was a major coup when a British privateer took a book of sea charts and sailing directions from a captured Spanish ship. The charts were soon copied and made available by William Hack, a London chart maker, who presented a set to James II in 1685 (Fig. 3). By then, systematic chart provision had begun elsewhere in Europe, initially with impetus from the Dutch and French trading companies rather than their navies, while commercial chart makers like Hack led the way in England. The possibility of finding better ways to determine longitude was bound up with this interest, as the poet John Dryden suggested in his historical poem Annus Mirabilis in 1667:

Fig. 3 – 'A description of the old town & the port of realejo' (El Viejo, Nicaragua), from 'A Waggoner of the South Sea', by William Hack, 1685, based on Spanish sea charts captured in 1680

What is longitude?

Latitude and longitude are the coordinates normally used to specify locations on Earth. The system was already established by the second century BC in the cartographic work of Hipparchus of Nicaea and enshrined by the second century AD in Ptolemy's *Geographia*, which described the mathematical concepts of a grid for mapping the world.

Latitude is the distance north or south of the Equator, measured as an angle from the centre of the Earth, and runs from 0° at the Equator to 90° at the North and South Poles. Each degree of latitude corresponds to roughly sixty nautical miles (111 km) on the Earth's surface. Lines of latitude run parallel to the Equator.

Longitude is the distance east or west, also measured as an angle from the Earth's centre. Lines of longitude, called meridians, run between the poles, where they converge. So, 1° of longitude on the Earth's surface is almost the same length as 1° of latitude at the Equator but diminishes to nothing at the poles. By convention, longitude is now measured from the Greenwich Meridian, and runs from 0° through Greenwich to 180° east and west on the other side of the globe. Until there was international agreement on this, whoever was measuring longitudes could choose any meridian or reference point they wished: Ptolemy, for example, used the island of Ferro (El Hierro) in the Canary Islands, as does the chart in Fig. 1, but London, Paris and many other places were used on different charts. Since it was difficult to measure with certainty, before the eighteenth century many charts did not show lines of longitude.

When plotting geographical positions, latitude and longitude are divided into degrees (°), minutes (') and seconds ("), with sixty minutes in a degree, and sixty seconds in a minute. The Empire State Building in New York, for example, lies at a latitude of about forty degrees, forty-four minutes and fifty-four seconds north of the Equator and at a longitude of about seventy-three degrees, fifty-nine minutes and ten seconds west of Greenwich. Its position is written as 40° 44' 54" N, 73° 59' 10" W.

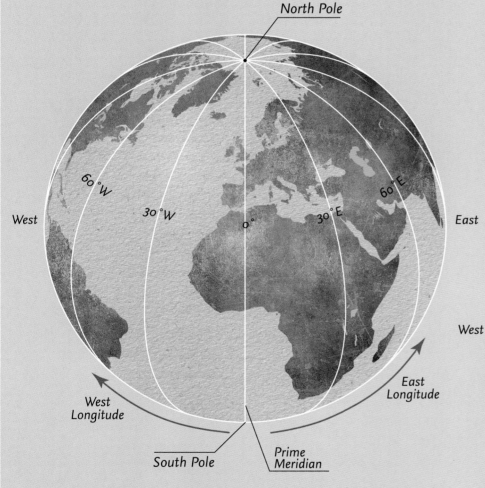

Fig. 4 – Longitude lines are imaginary lines on the Earth's surface that run from pole to pole around the globe and give the distance east or west from the Prime Meridian

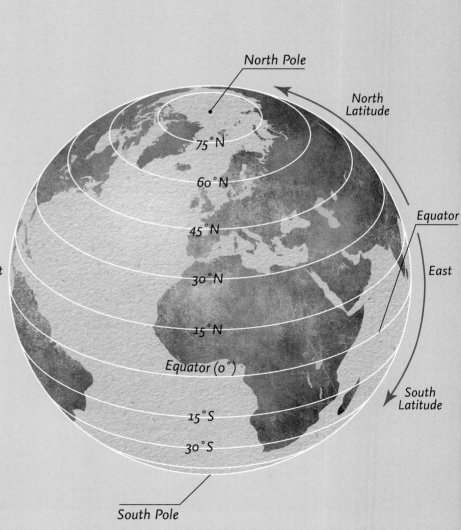

Fig. 5 – Latitude lines are imaginary lines on the Earth's surface. They run east and west around the globe and give the distance north or south of the Equator

Latitude relates to a definable reference (the Equator) and can be determined from the position of heavenly bodies such as the Sun or the Pole Star, but longitude is more difficult to determine because there are no natural references from which to measure. Since longitude is a distance in the direction of the Earth's daily rotation, the longitude difference between two places can be thought of as being directly related to the difference between their local times as defined by the Sun's position, local noon occurring when the Sun is highest in the sky. The Earth rotates through 360° in twenty-four hours, so one hour of time difference is equivalent to 15° of longitude; or, put another way, the Earth turns through 1° of longitude every four minutes.

Most longitude schemes were based on this principle and relied on an observer determining the time both where they were and, simultaneously, at a reference point with a known geographical position. The difficult part was knowing what time it was at the reference location. It was the same problem whether on land or sea, although a ship's movements made any observations much more difficult. Also, for marine navigation, the determination of longitude should not take so long that it became pointless, and any observations had to be possible on most days; that is, they could not be based on infrequent celestial phenomena.

There is another important issue related to this; as John Flamsteed (1646–1719), the first Astronomer Royal at Greenwich, noted in 1697, 'Tis in vain to talk of the Use of finding the Longitude at Sea, except you know the true Longitude and Latitude of the Port for which you are designed.'[4] In other words, navigators had to know exactly where their destination was and needed accurate charts on which to plot their position. So the story of finding longitude at sea is bound up with those of determining longitude on land and of creating better charts and maps.

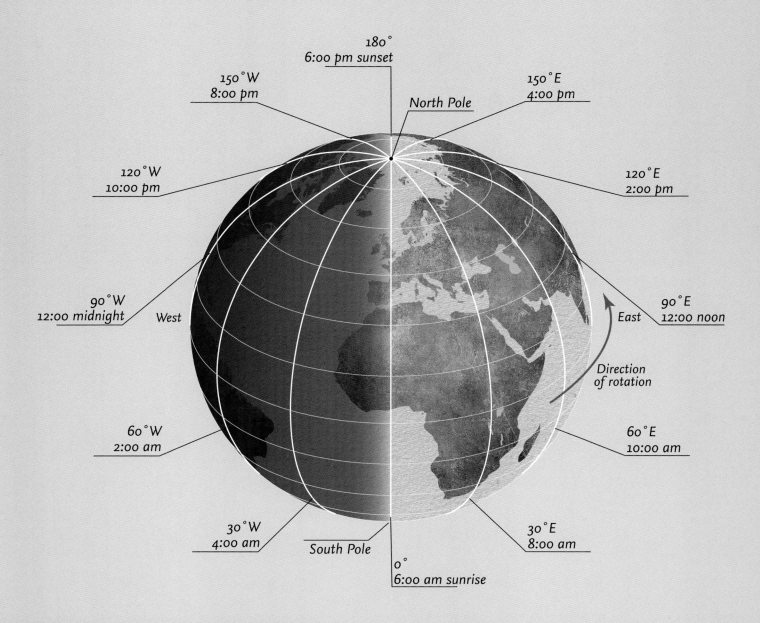

Fig. 6 – For places separated by 30° of longitude, the local time is two hours different – two hours later towards the east and two hours earlier towards the west. For places separated by 45° of longitude, the local times are three hours different

Instructed ships shall sail to quick Commerce;
By which remotest regions are alli'd:
Which makes one City of the Universe,
Where some may gain, and all may be suppli'd.[5]

Instruction, a footnote explained, was to be by 'a more exact knowledge of Longitudes'.

Though highlighted by Dryden and others, determining longitude was just one of many maritime problems for which solutions had long been sought. Seamen's health, including the control of scurvy, ensuring supplies of fresh water, understanding weather patterns, and building ships that were seaworthy, fast and, in the case of trading vessels, able to hold as much cargo as possible, would also tax the minds of seafarers, artisans and men of science for centuries to come. Yet it was longitude that would attract government attention.

The practice of navigation

Mariners were plying the oceans for centuries before the longitude 'problem' was solved. Many voyages were over relatively short distances and along familiar routes, often reasonably close to land, where being able to plot one's position with precision would have counted for little, but longer voyages often passed without incident too. This was because there were well-developed techniques for navigating a ship successfully that worked right across the globe.

Essentially, a mariner needed to know which way their ship was heading and how fast, where it had come from, where they were intending to go, how the sea and weather might affect them, and whether any hazards lay ahead. Tracking the ship's movements was the key and relied on the chip-log (or ship-log) for measuring speed in knots, and the magnetic compass for direction (Fig. 7). Throughout the voyage, the officers supervised regular observations of speed and direction, noting them down and later transferring the information to a written log (Fig. 8), together with wind direction and other remarks.

This information was used to fix the current position of the ship by plotting its direction and distance travelled from one point to the next – a procedure known as 'dead reckoning'. By applying the latest measurements to the previous day's position, and adjusting for the effects of wind and currents, the ship's navigator could plot the new position on the chart and note it in the written log. This was fairly straightforward over short distances. Over much longer distances, printed tables were used to convert the ship's various diagonal courses into changes of position north–south and east–west.

Latitude could be measured directly from the maximum height of the Sun or the Pole Star above the horizon. A range of instruments for these observations had been devised over the centuries, with those in use by the late seventeenth century including the cross-staff (see Chapter 2, Fig. 14) and, particularly among English sailors, the backstaff (Fig. 9). Each measured an angle between the celestial body (usually the Sun) and the horizon, from which latitude could be derived with a few simple calculations. Until the perfection of techniques described in later chapters, however, longitude could only be derived from dead reckoning, which was indispensable on long-distance voyages. On days when the weather allowed an astronomical observation for latitude, the difference between that and the latitude calculated by dead reckoning could be used to adjust the longitude estimate, hopefully improving its accuracy.

Constant vigilance was also essential – 'the best navigator is the best looker-out', Samuel Pepys noted.[6] This included watching for additional clues to check the ship's position, in particular when relatively close to the coast. Natural and man-made features, such as a headland, a church tower or a deliberately placed marker, were obvious signposts. As they headed 'north up the Yorkshire coast', for instance, Whitby sailors recalled that:

When Flamborough we pass by
Filey Brigg we mayn't come nigh
Scarborough Castle lies out to sea,
Whitby three points northerly.[7]

This local knowledge was also written down or published in books known as pilots or rutters (from the French *routiers*), which included descriptions and sketches of distinctive coastal features. When land was out of sight, birds, marine animals and plants could reveal its proximity and direction. On a voyage to

Fig. 7 – A mariner's compass made by Jonathan Eade in London, c.1750. The compass is mounted on gimbals to keep it steady on a moving ship. North is indicated by a fleur-de-lys

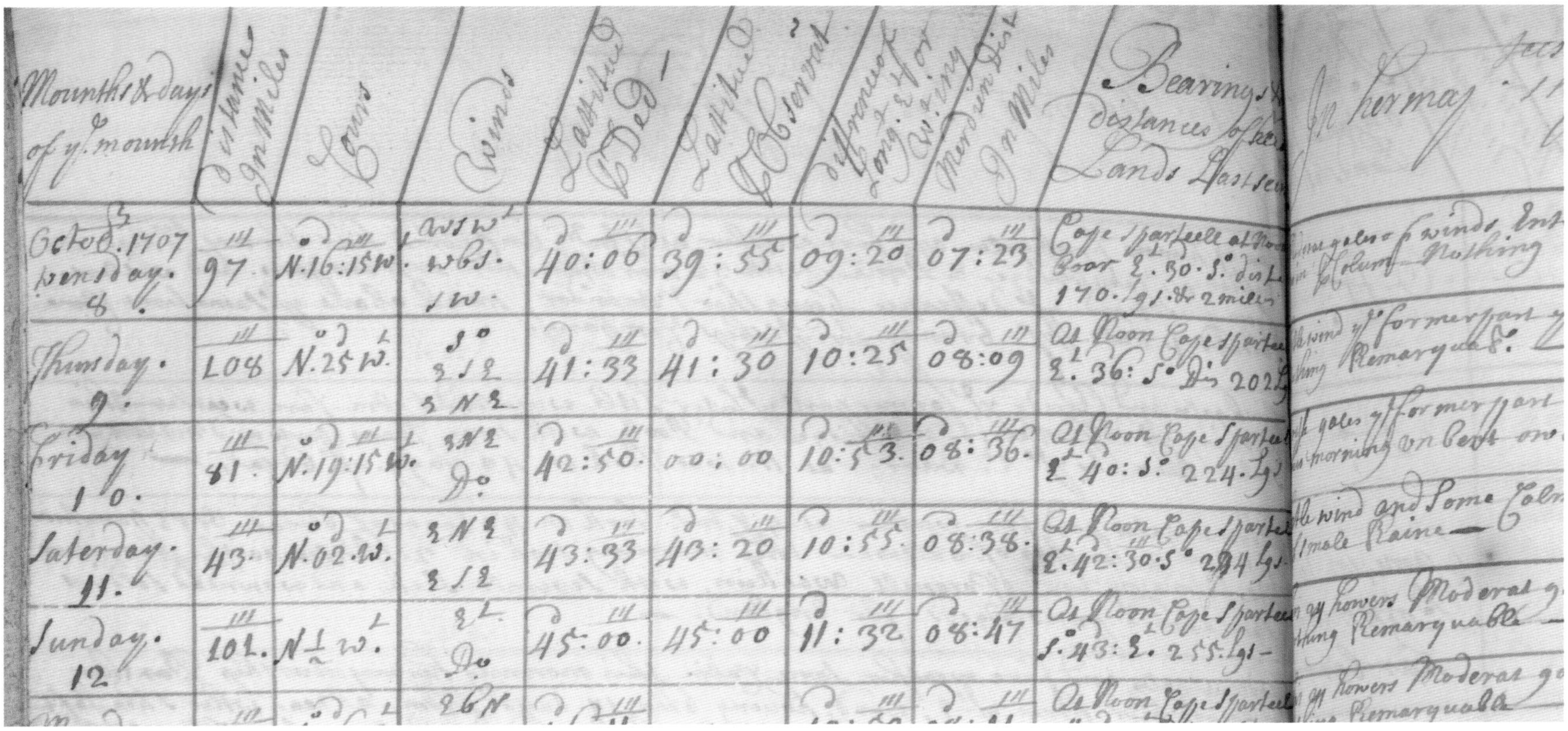

Fig. 8 – A page from the log of the *Orford* by Lieutenant Lochard, October 1707, showing the observations and results of calculations for latitude and longitude. There is also a column for general comments (detail)

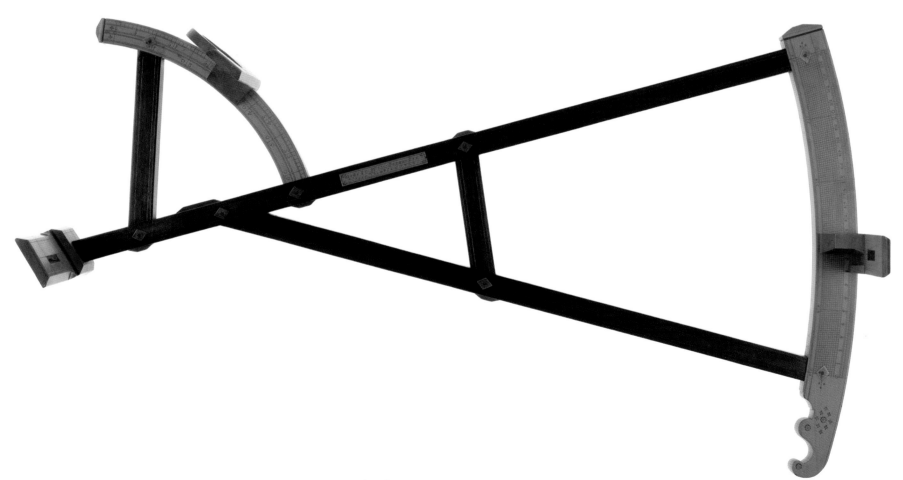

Fig. 9 – A backstaff, used to measure the angle between the Sun and the horizon; made of lignum vitae and boxwood by Will Garner, London, 1734

Fig. 10 – A seaman with a lead and line (right), from *The Great and Newly Enlarged Sea Atlas or Waterworld*, by Johannes van Keulen (Amsterdam, 1682) (detail)

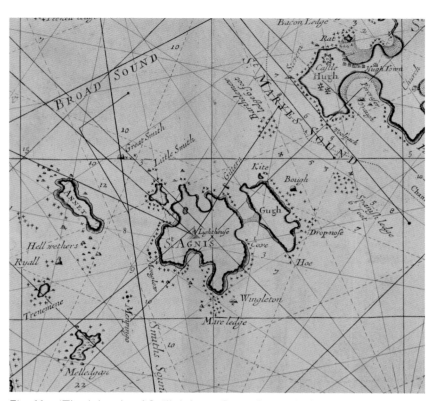

Fig. 11 – 'The Islands of Scilly', from *Great Britain's Coasting Pilot*, by Greenvile Collins (London, 1693). The lighthouse on St Agnis (St Agnes) was nine miles out of position (detail)

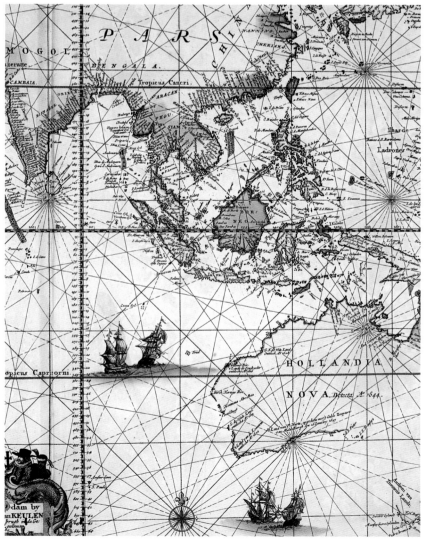

Fig. 12 – The Indian Ocean, from *The Great and Newly Enlarged Sea Atlas* (Amsterdam, 1682), showing Europeans' incomplete knowledge of the coastline of Hollandia Nova (Australia) (detail)

Fig. 13 – Navigation instruments used in the late seventeenth century, from *Practical Navigation*, by John Seller (London, 1672) (detail)

Philadelphia in 1726, Benjamin Franklin was reassured that they would soon arrive, having seen

> [an] abundance of grampuses, which are seldom far from land; but towards evening we had a more evident token, to wit, a little tired bird, something like a lark, came on board us, who certainly is an American, and 'tis likely was ashore this day.[8]

The lead and line (Fig. 10) – a lead weight attached to a long rope that was dropped at regular intervals to check the depth and nature of the seabed – gave further help. North Sea sailors, for example, boasted that they could tell west from east from the pebbles that came up with the lead (which had a hollow base 'armed' with tallow to pick up seabed samples): those in the west could be broken between one's teeth.

Lead, log and lookout worked well for coastal and short journeys, but might not be sufficient for longer ones. As European navigators embarked on increasingly ambitious voyages, often spending months in water too deep for sounding, they began to look to other methods for fixing their position. Being able to fix latitude and longitude with some degree of accuracy became more important.

One consequence of being unable to measure longitude directly was that seamen sensibly chose quite conservative routes. For example, if a ship set out on what the officers believed to be a direct course to its destination, there was the real danger that they would arrive at the correct latitude but find they had missed the destination. Unfortunately, they would not know whether they had sailed too far to the east or too far to the west, and so would not know which way to turn. The usual practice became to aim well to the east or west at the outset. Once the ship reached the latitude of their destination, they would 'run down the latitude' on a westerly or easterly heading, confident that landfall lay ahead. The buccaneer and explorer William Dampier (1651–1715) recorded using this method of latitude sailing on the *Batchelor's Delight* in 1684:

> we steered away N.W. by N. intending to run into the latitude of the Isles *Gallapagos*, and steer off West, because we did not know the certain distance, and therefore could not shape a direct Course to them. When we came within 40 minutes of the Equator we steer'd West ...[9]

It was a longer journey but they arrived safely a couple of weeks later.

On some routes, latitude sailing was a matter of safety. Approaching the south-west coast of India from the Cape of Good Hope, for example, trading vessels needed to avoid the dangerous waters near the Maldives and the Laccadive Islands (Lakshadweep). The recommended course was to keep west to a latitude of 8° or 9° North, where there were safe channels running east to the Indian coast. Ironically, the predictability of latitude sailing made it dangerous in wartime, when enemy ships simply waited at the appropriate latitude for victims to sail to them, a tactic employed by French privateers off the Windward Islands of the Caribbean.

Mariners' knowledge and skills, and the quality of their instruments, were crucial for effective navigation, as was the accuracy of charts and geographical data in printed manuals. However, these could be in error, even for areas close to home. Greenvile Collins's 1693 chart of the Isles of Scilly from his *Great Britain's Coasting Pilot* (Fig. 11), for example, placed the St Agnes lighthouse nine miles out of position, while the *Philosophical Transactions*, the Royal Society of London's journal, warned in 1700 that the information normally issued for ships heading into the English Channel was dangerously misleading. In less familiar waters, charts were likely to be even more unreliable or incomplete: it would not be until the nineteenth century that Australia's coastline would be fully drawn on European charts (Fig. 12).

Nonetheless, mariners had a set of methods that brought together centuries of accumulated seafaring knowledge with instruments and techniques that could be used to fix a ship's position and course, and navigate it safely from A to B and back again (Fig. 13). The staple was dead reckoning, the only routine method of determining longitude until the end of the eighteenth century, and the dominant one long after that. It was straightforward, used a relatively inexpensive suite of instruments and worked well enough in most situations.

Early attempts to measure longitude

While most mariners could not determine their longitude at sea with the tools normally available, there were occasional attempts to do so, since the theories were sound. The most obvious approach was to use eclipses, which were predictable and simultaneously visible from different locations. By comparing the local time of the eclipse on a ship with the predicted time at a specific place, noted in astronomical tables such as Regiomontanus' *Ephemerides* or Zacuto's *Almanach Perpetuum*, a mariner could work out the longitude difference from that place.

Eclipses had long been used for observations on land, including an ambitious project of the 1570s and 1580s to fix the positions of different parts of the Spanish empire and improve the maps and charts held secretly by the Council of the Indies, the governing body for the Spanish colonies in America. The scheme relied on local officials building a simple moondial and marking the position of the Moon's shadow on the dial when the eclipse began and when it ended. They then copied the marks onto paper and sent them with details of the length of the Sun's shadow at noon back to Spain for analysis. It was perfect for keeping sensitive cartographic information secret but the data was fiendishly complex to process and was riddled with error. A more successful project was Philipp Eckebrecht's world map of 1630, which used lunar eclipse data to plot many of the locations and was the first to equate one hour of time, astronomically determined, to 15° of longitude.

Eclipses could not provide a routine solution at sea, however, since they occur infrequently, although they could be tried out occasionally. Christopher Columbus made observations twice in the Caribbean, in 1494 and again in 1503, although his results were not impressive in terms of accuracy. That said, his observations were more in the way of experiments and were taken at anchor, rather than as part of routine navigation at sea, for which he used dead reckoning. Nonetheless, he did believe that 'with the perfecting of instruments and the equipment of vessels, those who are to traffic and trade with the discovered islands will have better knowledge'.[10]

Alternatively, eclipse observations from a ship could be compared with observations taken at another location, but only when the results could be brought together at a later date. On 29 October 1631, a Welsh explorer, Thomas James, viewed a lunar eclipse from Charlton Island in what is now Nunavut, Canada, during a voyage in search of the North-West Passage. Meanwhile, the mathematician Henry Gellibrand observed it at Gresham College, London, and was later able to calculate the longitude difference from James's figures as 79° 30' (a modern reckoning would place James's position as 79° 45' west of Gresham College). Gellibrand considered this an impressive result that augured well for future advances in the art of navigation.

Almost forty years later, John Wood used eclipses of the Moon for on-the-spot longitude determinations when he was master's mate on John Narbrough's 1669–71 expedition to the Pacific, which was instructed to bring back geographical information and lay the foundations for future trade in South America. Observations at sea of a partial eclipse on 26 March 1670 gave the longitude of Cape Blanco (Cabo Blanco in southern Argentina) as 69° 16' W, while today's value is 65° 45' W. Another observation a little further south on 18 September placed Port Desire (Puerto Deseado) at 73° W, the correct longitude being 65° 54' W. Wood also measured the position of the harbour of St Julian (Puerto San Julián, also in Argentina) from a conjunction of the Moon and Mars, calculating a longitude of 75° W, compared with a modern value of 67° 43' W. The observations showed significant errors by modern reckoning, not surprising given the instruments and data available, but they did demonstrate that determinations of longitude could be made while on expedition. While the infrequency of eclipses meant that they would never be routinely useful, other observations of the Moon had the potential to be used on a more regular basis and, as discussed in Chapter 2, some attempts to try them out were made in this early period.

Error and loss

Shipwrecks had many causes, just as they do today. Storms were a persistent problem but human error, including navigational mistakes, was also common. In many cases this was not simply about longitude determination but arose from a range of factors causing uncertainty as to a ship's position and surroundings. Without proper charts, no amount of position fixing could prevent disaster.

Fig. 14 (previous page) – Two English ships wrecked in a storm on a rocky coast, by Willem van de Velde the Younger, c.1700

Fig. 15 – *Sir Cloudisly Shovel in the Association with the Eagle, Rumney and the Firebrand, Lost on the Rocks of Scilly, October 22, 1707*

Problems could easily arise in relatively unfamiliar parts of the world, and might be compounded by hostile weather and unknown currents. This was something that William Dampier, the first person to circumnavigate the world on three separate occasions, discovered repeatedly in a turbulent seafaring life. Dampier ventured into the Pacific for the second time in 1703 in command of the *St George*, as part of an ill-fated privateering party with the *Cinque Ports*. As the ships rounded Cape Horn, storms hit with their expected venom. The *St George* attempted to crawl its way around the Cape but its position was soon uncertain as the winds took it wherever they wished. What happened next depends on whose account one believes. According to William Funnell, an officer whom Dampier later accused of desertion, Dampier ordered the ship north once he believed they were to the west of Cape Horn but two days later it turned out, 'contrary to all our expectations', that they were still five leagues east of Tierra del Fuego.[11] Dampier saw it differently. While he conceded that there was some uncertainty about their east–west position, sighting Tierra del Fuego was not so unexpected:

> for it is well known the Evening before, I told them we should see Land the next Morning, *that of Terra del Fuego, the South Part of it: Now I look upon that to be a greater Mistake, to take one side of the Land for the other, than 'tis to be mistaken that we were Westward of the whole Island, and miss his Longitude* ...[12]

In any case, they were forced to brave the Horn once more.

Their troubles did not end there. Having separated from the *Cinque Ports*, the *St George* headed north towards the Juan Fernandez archipelago, off the coast of Chile, which was a regular rendezvous and watering spot for ships entering the Pacific. The typical approach to Juan Fernandez was to run down the latitude from the coast of Chile but, according to Funnell, the *St George* sailed right past because Dampier failed to recognize the islands. They finally returned after three days without sight of land, only to find the *Cinque Ports* safely anchored there. Dampier's information and memory had led him astray. Incidentally, one of the sailors on board the *Cinque Ports* was Alexander Selkirk, the inspiration for Daniel Defoe's *Robinson Crusoe*, who decided he would rather be abandoned alone on an uninhabited island in the archipelago than remain on the unseaworthy *Cinque Ports*. Despite the privations of life on the island, it proved to be wise decision as the *Cinque Ports* foundered later in its journey.

Mariners feared Cape Horn with reason, and the same was true of the western coast of Australia, which is littered with offshore reefs and islands that saw the demise of many ships plying their trade between Europe and Asia. The most notorious incident followed the loss of the VOC ship *Batavia* on its maiden voyage. *Batavia* sailed from the Netherlands in October 1628 and was in the southern Indian Ocean eight months later, heading along the recommended route eastwards before turning north for Java once it reached the correct longitude. By 4 June, it was approaching the Houtman Abrolhos, a known hazard off the west coast of Australia named ten years earlier by Frederik de Houtman, from the Portuguese *abre os olhos*, meaning 'open your eyes'. The *Batavia*'s pilot knew he was approaching the reefs but seems to have ignored the danger signs and the ship struck. Of 322 on board, forty drowned during the shipwreck, and more than 110 men, women and children were killed as they awaited rescue in a tale of mutiny and murder that made for sensationalist reading back in Europe.

Remote, unknown waters presented obvious dangers but there were plenty closer to home as well. Indeed, the Royal Navy's worst maritime disaster of this period occurred not hundreds or thousands of miles away, but off the Isles of Scilly. Having concluded naval operations in the Mediterranean, a fleet under the command of Admiral Sir Cloudesley Shovell set sail for England at the end of September 1707. It should have been a routine voyage in well-known waters, even though they hit gales as they headed home. Just over three weeks in, the Admiral ordered his ships to heave to and check their position, concluding that they were safely

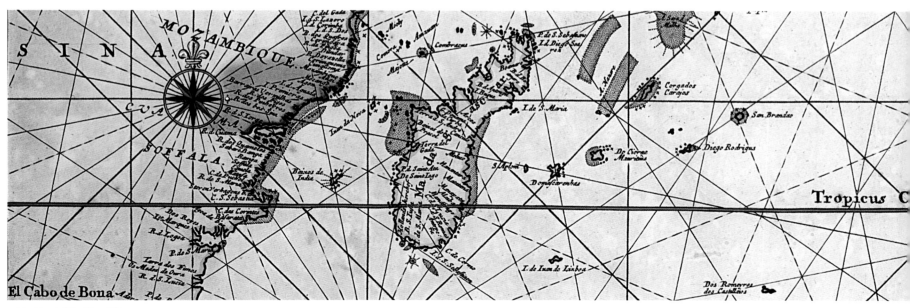

see page 24, Fig. 12

in the English Channel. On the evening of 22 October, however, five ships struck the outlying rocks of Scilly (Fig. 15). Within hours four had sunk and at least 1600 men, including Shovell, were dead. It was a national tragedy.

Many causes have been cited: weather; tides and currents; compass error; even longitude. What the surviving log-books show is that variable navigational abilities and unreliable data were the main culprits. The officers' latitude determinations from backstaff observations, for instance, had an average spread of 25.5 nautical miles (47.2 km), those by dead reckoning a spread of 73 nautical miles (135.2 km). More dangerously, their geographical data was flawed. Most of the fleet took Cape Spartel at the entrance to the Strait of Gibraltar as their point of departure. Its latitude and longitude were listed in manuals such as Colson's *New Seaman's Kalendar* and Seller's *Practical Navigation*, but their figures varied widely. Combined with inaccurate charts (see Fig. 11) and generally moderate navigational skills, poor data landed the fleet in its perilous position.

The drive to improve navigational knowledge

The known hazards of the sea and the resulting losses were a spur to improve all aspects of seafaring, including navigation. Among Europe's maritime states, possible improvements were of obvious interest to those wielding commercial and political power as they sought to strengthen naval and trading operations. As will be seen in Chapter 2, the various rewards offered for longitude solutions from the sixteenth century onwards, and the foundation of state observatories in the seventeenth, arose within this context.

Identical motives lay behind initiatives to improve navigational training. State-backed schools to train and license navigators and pilots engaged in long-distance trade were founded in Spain and Portugal in the fifteenth and sixteenth centuries, and these inspired a number of French navigation schools in the second half of the seventeenth century. Britain was not far behind its rivals when, in 1672, Samuel Pepys, by then Clerk of the Acts of the Navy, led moves to create 'a Nursery ... of Children to be educated in Mathematicks for the particular Use and Service of Navigacon'.[13] Granted a charter by Charles II, the Royal Mathematical School at Christ's Hospital took in forty boys each year, from 1673, to study mathematics and navigation to prepare them for life in the merchant service or Royal Navy.

Significantly, the school had the support of Isaac Newton (1642–1727) and the astronomers John Flamsteed and Edmond Halley (1656–1742), who saw this as a way in which their work could have tangible public benefit. Flamsteed, who taught some of the boys at the Royal Observatory, wrote to Pepys of the school's value, foreseeing a time when trained seamen would fix longitudes from astronomical observations, 'whereby the faults of our present Mapps and Sea Charts ... will be corrected and a halfe the Business of navigation perfected'.[14] Another school initiated in 1712 as part of Greenwich Hospital for Seamen followed similar lines, with pupils first taught (from 1715) by Thomas Weston, assistant to the Astronomer Royal.

Pepys continued his campaign to improve standards after becoming Secretary of the Admiralty. Having observed naval navigation in action, his assessment was scathing:

> it is most plain from the confusion all these people are to be in how to make good their accounts (even each man's with itself) and the nonsensical arguments they would make use of to do it, and disorder they are in about it ... that it is by God Almighty's providence and great chance and the wideness of the sea that there are not a great many [more] misfortunes and ill chances in Navigation ...[15]

The school solution had been one way to address the problem. Pepys anticipated that, with concerted state support and the help of astronomers and mathematicians, the painful situation he described might be alleviated to the benefit of the nation.

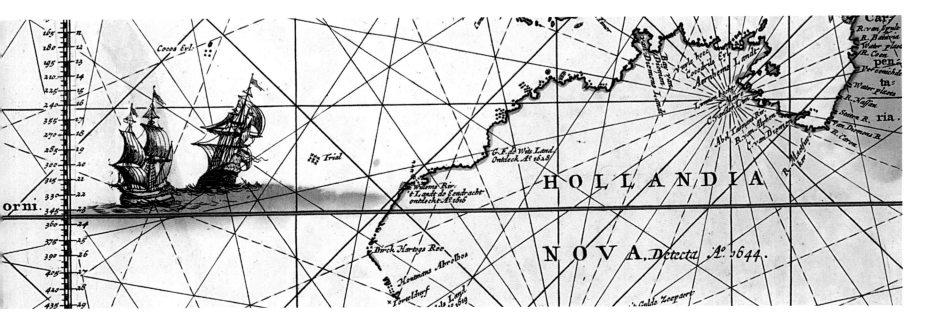

Up until the eighteenth century, longitude determination at sea was one of a number of challenges that faced naval and merchant fleets worldwide. As different nations became the dominant players in maritime affairs, so their political and commercial leaders were willing to give encouragement to anyone offering to solve any of the myriad problems that diminished profits and put lives in danger.

In Britain, the drive to improve navigational knowledge gained impetus as a result of the loss of Admiral Sir Cloudesley Shovell and his men. As an experienced and high-ranking naval officer who knew the dangers only too well, Shovell had had an interest in navigational improvements; for instance, he had met with Isaac Newton in 1699 to examine a proposal for finding longitude put forward by a Monsieur Burden. His death in 1707 alongside so many of his men was a national disaster and, though not solely (if at all)

attributable to problems in determining longitude, would be cited in lobbying for a Longitude Act seven years later.

Since there were techniques allowing mariners to navigate with some confidence, one could argue that the measurement of longitude was not an insurmountable problem for them. Nonetheless, it was a practical issue whose solution was felt by some to be within reach and of obvious benefit. For mathematicians, astronomers and cartographers, in particular, it was an intellectual challenge and a practical conundrum in which the peculiarities of being at sea merely hindered elegant mathematical and astronomical solutions. To their way of thinking, here was an arena in which their skills might be called upon in the service of humanity, perhaps earning them fame, fortune and influence in the process. Longitude was a problem for which they believed they might have the answer, and it was they who would put it on the political agenda.

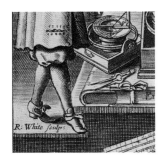    

CHAPTER 2

The Contenders

in the Judgment of Able Mathematicians and Navigators, several Methods have already been Discovered, true in Theory, though very Difficult in Practice ...

'An Act for Providing a Publick Reward for such Person or Persons as shall Discover the Longitude at Sea' (1714)[1]

Longitude, as a definable problem that could be separated out from the myriad risks and uncertainties of maritime travel, was of interest to theorists as well as practical navigators. It was, in fact, an issue that advocates of natural and experimental philosophy – what we today call science – latched onto as one for which their approach might be particularly successful. It was clear that finding a solution would be a propaganda coup for the new scientific institutions. It would be irrefutable proof that experimental, observational and mathematical methods, overseen by gentleman philosophers, could be applied to practical issues of importance to national interests.

Within Britain, maritime matters had been a significant focus for the Royal Society of London. That these included tackling longitude was a matter for satire, as in this anonymous poem, 'In praise of the Choyce company of Philosophers and witts who meet on Wednesdays weekley, at Gresham Colledge', written in 1661:

> The Colledge will the whole world measure,
> Which most impossible conclude,
> And Navigation make a pleasure
> By finding out the longitude.
> Every Tarpalling shall then with ease
> Sayle any ships to th' Antipodes.[2]

Fellows of the Royal Society were to play a significant role in the passing of the Longitude Act by the British Parliament in 1714, which was itself to transform the relationship between scientific expertise and the state. It was, however, just the latest in a long line of initiatives to reward anyone able to arrive at workable solutions to the problem.

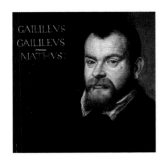

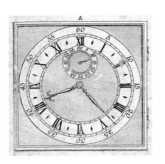

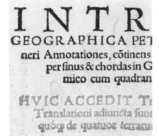

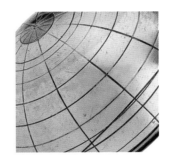

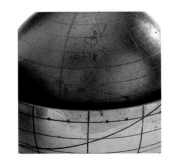

Longitude rewards

Rulers and states with maritime ambitions who were convinced that the longitude problem could be solved looked for ways to hasten solutions. Their conviction was, it seems, more often the result of lobbying by those likely to gain from financial incentives than a response to calls for assistance from practical seamen. The first such incentive scheme was established in Spain, the leading maritime power of the sixteenth century, by Philip II in 1567. This was followed in 1598 by Philip III's offer of a reward of 6000 ducats, plus 2000 a year for life – some sixty times the annual income of a labourer – and 1000 towards expenses. The large reward was never paid out but several promising inventions were recognized by the repayment of expenses.

The life-changing size of the rewards on offer, and the fact that workable and complete solutions to the problem were clearly not appearing, led to the whole enterprise being satirized. Cervantes, author of *Don Quixote*, was one among many over the centuries to mock those who were mad enough to attempt such an impossible task, or who drove themselves mad in its pursuit. He wrote in 1613 of a mathematician who found fixing the longitude like chasing a will-o'-the-wisp:

> Two and twenty years I have been employed in finding out the longitude ... and imagining oftentimes, that I have found it, and that it cannot possibly escape me, when I do not in the least suspect it, I find myself as far to seek as ever, which fills me with surprise and astonishment: it is the same with the square of the circle, which I have come so nigh finding out, that I know not, nor can imagine why I have it not at this time in my pocket ...[3]

Finding the longitude – like squaring the circle, creating perpetual motion or predicting the future – was often presented as a fool's errand. Yet the solutions were, in theory, well understood. What they needed were technical responses to the various challenges thrown up by imperfect astronomical tables and conditions at sea. The answer might, therefore, have seemed tantalizingly close. To nudge them forward, the States General and States of Holland and, less officially, a number of other governments and individuals followed the Spanish example by offering rewards for meeting these challenges. Serious and important ideas were submitted, trials were made and some money paid out.

The Dutch scheme began in 1600 and established a range of rewards, with large one-off sums, annuities and smaller sums for those with promising ideas ready for trial. The top rewards increased over time, with the States of Holland offering 3000 guilders in 1601 and 50,000 in 1738. These sums were sufficiently alluring to attract a steady flow of ideas. There were forty-six submissions between 1600 and 1775, judged by ad hoc committees of theoretical and practical experts, including surveyors, teachers of navigation and university professors. The Dutch East India Company took an interest in the process, reflecting the particular risks and rewards of their trade routes. As with the Spanish reward, a whole range of ideas and solutions was offered, and a number underwent sea trial.

England and France came relatively late to longitude research but their arrival coincided with a period of significant advances in astronomy and instrument-making. Government and royal interest in finding longitude solutions led in both nations to the patronage of individuals with plausible methods, to the theme being taken up by the learned academies – the Royal Society of London and French Académie des Sciences, founded in 1660 and 1666 respectively – and to the establishment of observatories in Paris in 1667 and at Greenwich in 1675.

In terms of both the legislation and the work it was meant to encourage, the passing of the British Longitude Act of 1714 was more a case of continuity than of change. It did, however, come at a time when key successes and initiatives relating to navigation and longitude determination were bearing fruit: several areas of research, with long and sometimes distinguished track records, were earmarked as most likely to lead to success. Given advances in these areas over the preceding decades, there was probably some confidence that at least partial success could be achieved in the not-too-distant future.

The Longitude Act of 1714

If legislators believed that success was imminent, they were encouraged in that view by lobbyists and scientific experts. Their call for a British scheme to echo the Spanish and Dutch was answered on 9 July 1714, when Parliament passed 'An Act for Providing a Publick Reward for such Person or Persons as shall Discover the Longitude at Sea'. It provided

> That the First Author or Authors, Discoverer or Discoverers of any such Method ... shall be Entitled to ... a Reward, or Sum of ten thousand Pounds, if it Determines the said Longitude to One Degree of a great Circle, or Sixty Geographical Miles; to fifteen thousand Pounds, if it Determines the same to Two Thirds of that Distance; and to Twenty thousand Pounds, if it Determines the same to One half of the same Distance ...[4]

The value of this reward, in today's terms, depends on the yardstick. £20,000 in 1710 can be calculated as worth over £1.5 million early in the twenty-first century, or as a sum that would have paid the annual wages of more than 600 craftsmen in the building trade or bought nearly 5000 cows.[5] By any measure it was a large amount of money but the Act asked for a great deal: the successful method was to be 'Tried and found Practicable and Useful at Sea'.

The emphasis on practical utility was underlined by the provision that half of the reward could be given to a method that ensured 'the Security of Ships within Eighty Geographical Miles of the Shores, which are the Places of the greatest Danger'.

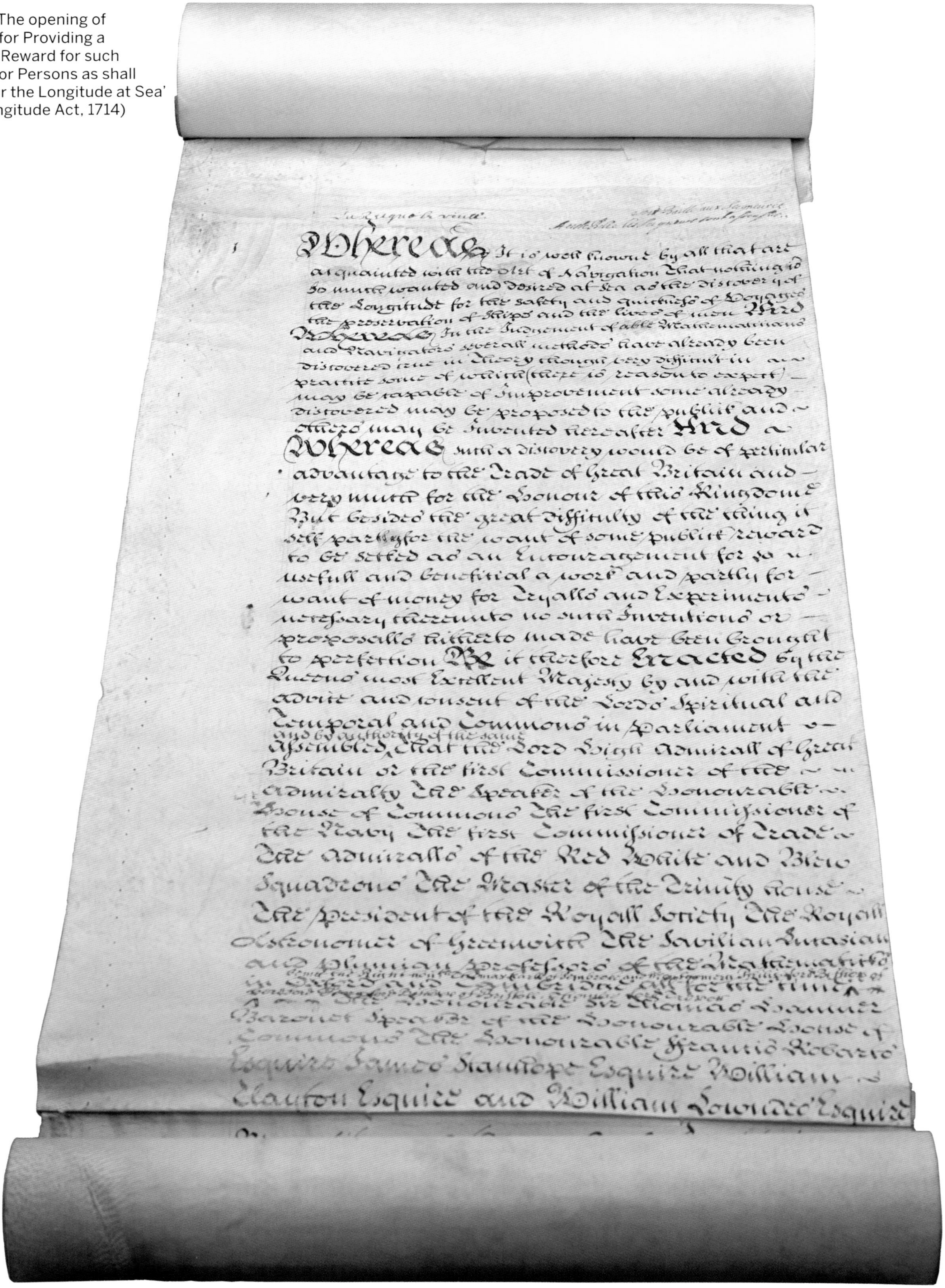

Fig. 1 – The opening of 'An Act for Providing a Publick Reward for such Person or Persons as shall Discover the Longitude at Sea' (the Longitude Act, 1714)

Fig. 2 – Isaac Newton, by Charles Jervas, 1717

However, the full reward was dependent on a successful long-distance sea trial:

> when a Ship by the Appointment of the said Commissioners, or the major part of them, shall thereby actually Sail over the Ocean, from Great Britain to any such Port in the West-Indies, as those Commissioners, or the major part of them, shall Choose or Nominate for the Experiment, without Losing their Longitude beyond the Limits before mentioned.

The Act also allowed for the advancement of sums up to £2000 'to make Experiment' of promising schemes. This appears to have meant costs associated with the trial but was later interpreted more broadly by the Commissioners. It was also possible to give rewards to schemes that failed to achieve the desired accuracy, but were nevertheless thought to be 'of considerable Use to the Publick'.

Perhaps most importantly, the Act appointed the Commissioners of Longitude. They were the Lord High Admiral, or First Commissioner of the Admiralty; the Speaker of the House of Commons; the First Commissioner of the Navy; the First Commissioner of Trade; the Admirals of the Red, White and Blue Squadrons of the Navy; the Master of Trinity House; the President of the Royal Society; the Astronomer Royal; the Savilian, Lucasian and Plumian Professors of Mathematics (the first at Oxford University, the other two at Cambridge); and ten named Members of Parliament. They were not yet a standing body or a 'Board', but the Act was significant in creating a group, mostly selected because they held another relevant office, that combined political, navigational and scientific interests. While the President of the Royal Society and university professors had previously acted as advisors, this Act brought philosophers and mathematicians directly into a process of allocating government funds.

When action was being considered in June 1714, it was one of the future Commissioners, Isaac Newton, who presented evidence to a Committee of the House of Commons. He was then President of the Royal Society and, as such, an unofficial advisor to politicians on matters scientific (Fig. 2). He was also working directly on the mathematical and astronomical theories that had potential to help solve the problem. Newton's written statement noted, in a phrase echoed in the legislation, that 'there have been several Projects, true in theory but difficult to execute'. He provided a succinct summary of the methods and the problems still surrounding them:

> One is by a Watch to keep Time exactly. But, by reason of the motion of a ship, the variation of heat & cold, & the difference of gravity in different Latitudes, such a Watch hath not yet been made.
> Another is by the Eclipses of Jupiter's Satellites. But by reason of the length of Telescopes requisite to observe them & the motion of a ship at sea, those Eclipses cannot yet be there observed.
> A third is by the place of the Moon. But her Theory is not yet exact enough for this purpose. It is exact enough to determine her Longitude within two or three degrees, but not within a degree.
> A Fourth is Mr Ditton's project. And this is rather for keeping an account of the Longitude at sea than for finding it if at any time it should be lost, as it may easily be in cloudy weather ...[6]

The first three methods on this list were, to Newton, the most promising. All were a means of carrying or finding a reference time against which to compare observations of local time on board ship. If practicable, they would allow difference in longitude to be established anywhere in the world. The last method, 'Mr Ditton's project', as Newton said, targeted particular circumstances rather than being a universal solution to the problem. It is, however, of particular interest to the story, being the cause of parliamentary interest in the issue of longitude in 1713.

'Mr Ditton's project'

The Mr Ditton to whom Newton referred was Humphry Ditton (1675–1715), Master of the Royal Mathematical School at Christ's Hospital. In reality, it was Mr Ditton's and Mr Whiston's project, but the latter was an individual with whom Newton, once close, had now broken ties. Perhaps Newton could not quite bring himself to write the name, despite the presence of William Whiston (1667–1752, Fig. 3) at the parliamentary committee. Whiston had been Newton's chosen successor as Lucasian Professor of Mathematics at Cambridge in 1702, but was expelled from the university in 1710 for his unorthodox theological views. Those views were largely shared by Newton but he was anxious to avoid a public accusation of heresy. Whiston blamed their rupture on Newton's 'fearful, cautious, and suspicious Temper'.[7]

After 1710, Whiston made his living as a scientific and theological lecturer, author and, he hoped, longitude projector – that is, someone who sought backing for a scheme, or project, intended to solve the longitude problem. He had lectured with Ditton from at least 1712 and they were promoting their longitude scheme the following year, through newspaper advertisements and letters drumming up English support by asking questions about foreign longitude rewards. In 1714, two petitions to Parliament appeared: one in April from Whiston and Ditton, and another in May recorded as being from 'several Captains of her Majesty's Ships, Merchants of *London*, and Commanders of Merchant-men'. The latter petition suggested that public 'Encouragement' would aid the search for a longitude solution.[8] It was this, in which Whiston may also have had a hand, that instigated the parliamentary committee at which Newton presented his evidence. Prominent Whig politicians, whose patronage Whiston enjoyed, were to steer the new legislation.

Claiming inspiration from the extraordinary display of fireworks on the Thames that celebrated the end of the War of the

Fig. 3 – William Whiston, by an unknown artist, c.1690

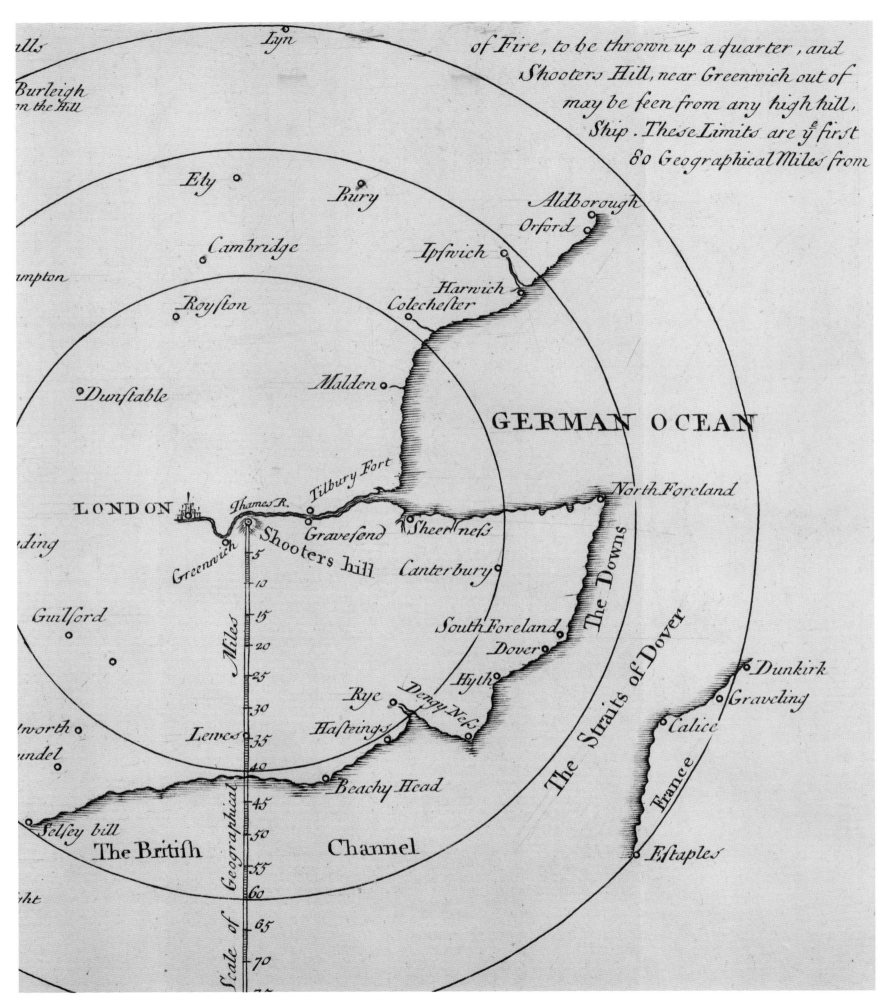

Fig. 4 – The limits of viewing the flash from a mortar fired at Shooter's Hill, near Greenwich, from William Whiston's *The Longitude Discovered* (London, 1738) (detail)

Fig. 5 – A terrella (or 'little earth'), a spherical lodestone used to model the Earth's magnetic field, c.1600

Spanish Succession in 1713, Whiston and Ditton proposed that vessels moored at known locations could fire shells vertically to 6440 feet at set times. Navigators would keep a look out for the lights and gauge their bearing and distance relative to the moored vessel by compass and by timing the difference between the flash and sound of the shell, or by measuring its elevation. Whiston and Ditton thought large shells might be visible for a hundred miles and that, where deep seas meant hulks could not be moored, ships might run down the latitude until they neared the next signal post. Newton seemed disinclined to comment more than necessary: 'How far this is practicable & with what charge, they that are skilled in sea affairs are best able to judge'.⁹

Whiston and Ditton presented their scheme more fully in a pamphlet addressed to the newly appointed Commissioners of Longitude. They offered it as a practical idea that, without universally solving the problem, would make a material difference. It was 'easy to be understood and practis'd by Ordinary seamen, without the Necessity of any puzzling Calculations in Astronomy' but would 'prevent the Loss of abundance of Ships and Lives of Men'. The signals could provide both latitude and longitude, might be used to give exceptional warnings of bad weather, and would have most success in the areas of greatest danger – that is, near coasts. Here they invoked the maritime tragedy that occurred off Scilly in 1707, claiming their scheme 'would certainly have sav'd all Sir *Cloudsly Shovel's* Fleet, had it been then put in Practice'.

To put the idea into effect, Whiston, who continued the project after Ditton's untimely death in 1715, relied on the skills of London's firework makers and gunners as he began trials on Hampstead Heath and Blackheath (Fig. 4). There was merit in the idea, and explosives were later occasionally used to measure distances in survey work, but there were serious practical problems. Not least was the difficulty of mooring vessels in deep water, despite a claim that anchors might be secured by reaching down to supposedly still layers of water far below the surface.

It was all too easy for the firework scheme to be ridiculed, particularly by Tory satirists, who connected it to Whiston's fiery and suspect theology, and cast the whole concept of longitude rewards as a Whig folly. Nevertheless, projecting, publicizing and finding patronage for longitude schemes remained one of Whiston's major activities and sources of income. He explored all the accepted avenues of research, including one not mentioned by Newton: the Earth's magnetism.

Magnetic variation and inclination

The idea that patterns in the Earth's magnetic field might be a means of fixing position at sea had a long history and continued to be investigated in the eighteenth century and even into the nineteenth. It is interesting that Newton did not mention it in his evidence to the parliamentary committee, especially since he was joined there by Edmond Halley, an experienced astronomer, mathematician and navigator who had investigated these phenomena himself. As with Whiston's signals, this method was about finding position relative to known locations rather than finding longitude itself. Presumably Newton and Halley therefore considered it discounted as a universal solution.

Most of the incoming proposals to the Spanish and Dutch longitude reward schemes were based on patterns in the Earth's magnetic field, and many would be put to the British Commissioners of Longitude. There were two patterns that were investigated, with the hope that they were regular enough to be mapped and used. One was magnetic variation (also known as magnetic declination), which is the angular difference between magnetic north, shown on the compass, and true north, determined by the Sun or stars. A positive variation shows that magnetic north is east of true north, a negative one that it is to the west. The other was magnetic inclination, or magnetic dip, which is measured by the compass needle's vertical rather than horizontal deviation. This is caused by the needle aligning itself with the Earth's curving lines of magnetic force.

These were phenomena that had long been observed and investigated: variation had to be understood by navigators to correct steering directions, if not for position finding. In trying to make sense of the patterns of observational data, natural philosophers attempted to describe the Earth as, or containing, a giant magnet. One of the most famous of these accounts was *De Magnete*, published by William Gilbert, a London physician, in 1600. He undertook much of his research with spherical lodestones: known as terrellae, meaning 'little earths', these magnetic rocks were used to model patterns of geomagnetism (Fig. 5).

Magnetic inclination was also explored as a means of finding latitude, which, given that the Earth's lines of magnetic force run north–south, had some plausibility. However, experiments had shown that the variations were too irregular and the observations

Fig. 6 – An amplitude compass, used for measuring magnetic variation from the apparent bearing of the Sun's rising or setting; made by Ferreira, Lisbon, 1780

Fig. 7 – Edmond Halley, by Thomas Murray, c.1690

too difficult to make at sea. In any case, astronomical observations were becoming much more effective for determining latitude. Thus, while schemes relating to magnetic inclination for latitude or longitude did not disappear, and even occasionally recurred in navigational textbooks, it was magnetic variation that had more impact. It involved significantly easier on-board observations and had a more plausible theoretical underpinning. This too was challenged, however, when it was demonstrated that the patterns change over time as well as place.

Nevertheless, navigation by magnetic variation was actually achieved, albeit in restricted locations or on familiar routes. The necessary tools were an amplitude compass (Fig. 6), to measure variation from observations of the sun and a chart recording previously observed lines of equal variation, against which to plot the ship's position. This could be effective in specific circumstances, where the charting was detailed, and the lines ran nearly north–south and were reasonably close together. Some Portuguese navigators, for example, and those on Dutch East Indiamen, put the method into practice at various times during the seventeenth and eighteenth centuries, many apparently satisfied with the results.

Research into magnetism was a serious interest at the Royal Society, with demonstrations by their curator of experiments, Robert Hooke (1635–1703), who developed his own magnetic theory. Between 1668 and 1716, annual predictions by Henry Bond, a teacher of mathematics and navigation, were published in the Society's journal, with the aim of encouraging magnetic observations against which his theory might be tested. Bond's claims were investigated by a Royal Commission in 1674 and, although there was some doubt, he was paid £50 and given licence to publish his book *The Longitude Found*. On the basis that the six Commissioners were all Fellows, the book claimed the Royal Society's approval, to which its President, Viscount Brouncker, objected strongly.

The Society's interest nevertheless continued and was instrumental in persuading the government and the Navy to fund and equip a scientific voyage that would, among other things, chart magnetic variation as widely as possible. Edmond Halley (Fig. 7) was, very unusually for a civilian, given command of a specially built naval vessel, the *Paramore*, and set sail on two voyages in 1698 and 1699. He published charts of magnetic variation in 1701 and 1702 (Fig. 8), noting that they might be useful both for rectifying courses where compass readings might be unreliable, and for estimating longitude in places where the lines of similar variation were almost parallel to a meridian, provided that such charts were kept up to date to reflect change of variation over time.

While Halley did not produce updated charts, others did and they were put to use. However, the fact that this inexact, localized, practice-based and changeable method was not mentioned at the 1714 parliamentary committee underlines Newton's view at the time that Parliament should be looking for a more complete solution. While Whiston and Ditton's scheme had to be mentioned – and, by Whiston's account, Newton's initial ignoring of it risked the complete rejection of the proposed legislation – it perhaps served as a contrast to the great aim of finding a method that could be applied confidently at any location. It was, Newton suggested, only the astronomical and timekeeper solutions that held out that promise.

'the Eclipses of Jupiter's Satellites'

While rocket signals and magnetic schemes were about finding a means of fixing position relative to a known location (a moored hulk or charted magnetic feature), the other methods focused on the long-understood relationship between time and longitude. These were the only universal solutions to finding longitude at sea and, as Newton explained more than once to the Admiralty, only astronomical methods could be used to find longitude if it had been lost. The downside of astronomy was that observations could usually only be done at night, and sometimes irregularly if the target object was not in the right position, which meant that dead reckoning and other techniques were still required. Observations could also be also hampered by clouds, although this was equally true for calculating latitude and local time, without which neither astronomical nor timekeeper methods were effective.

The use of lunar and solar eclipses was the earliest of several potential astronomical methods for finding longitude. One key branch of research was to find ways of using the Moon's position on a more regular basis, while the discovery of the moons (satellites) of Jupiter, with much more frequent eclipses, opened up new opportunities. The satellites were discovered in 1610 by Galileo Galilei (1564–1642, Fig. 9) with the use of a new instrument – the telescope. It revealed that Jupiter was orbited by four satellites that would disappear and reappear with useful regularity as they passed in front of or behind the planet (Fig. 10). They provided, in essence, a celestial timekeeper, visible at the same time from different points on Earth.

Galileo quickly realized that this was a potential means of finding longitude and, having drawn up provisional tables to predict the satellites' motions, he attempted in 1616 to interest the Spanish in a proposal to make 100 telescopes and teach navigators the method. Having failed to convince them, he began protracted negotiations with the Dutch government in 1636, which only ended with his death six years later. The scepticism of the Spanish longitude committee was undoubtedly related to the practicalities of the method. Observing objects as small as Jupiter's satellites with a telescope from a moving ship was clearly going to be very difficult. Even a century later, when the production of telescopes and lenses had vastly improved, Newton noted that 'by reason of the length of Telescopes requisite to observe them & the motion of a ship at sea, those Eclipses cannot yet be there observed'.[6]

Galileo recognized this problem and looked for a means of steadying the observer. He designed a helmet, the *celatone*,

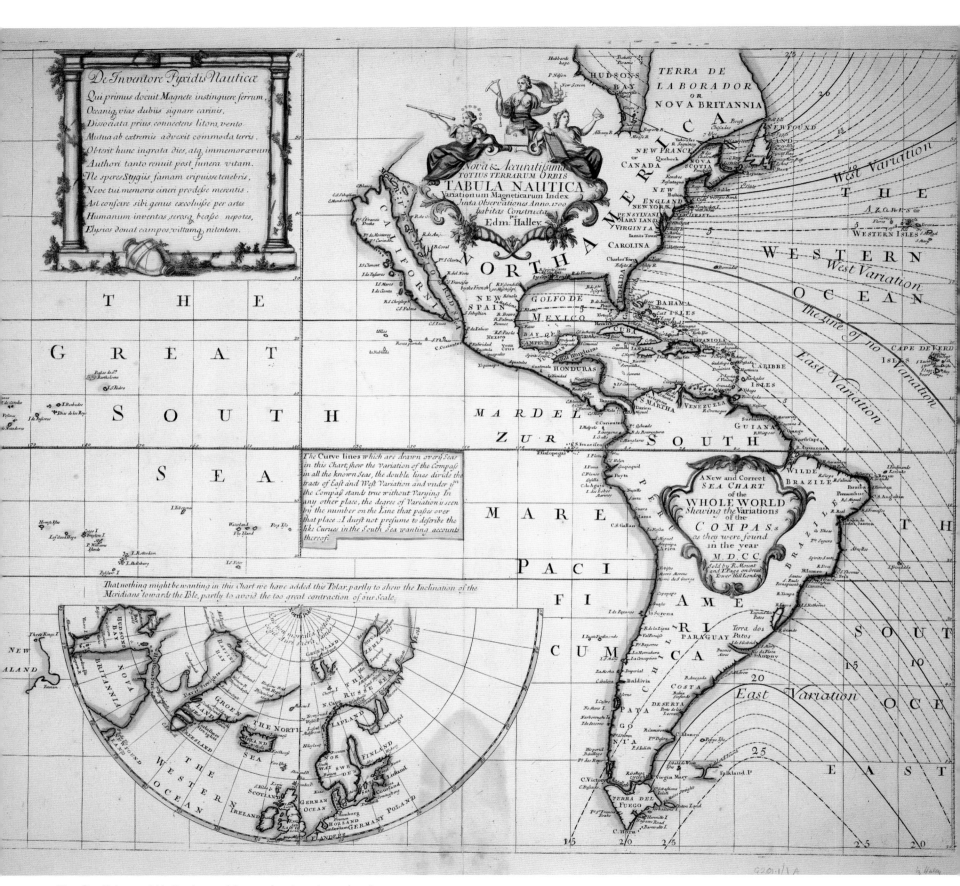

Fig. 8 – Edmond Halley's world sea chart on two sheets, showing lines of equal magnetic variation, 1702

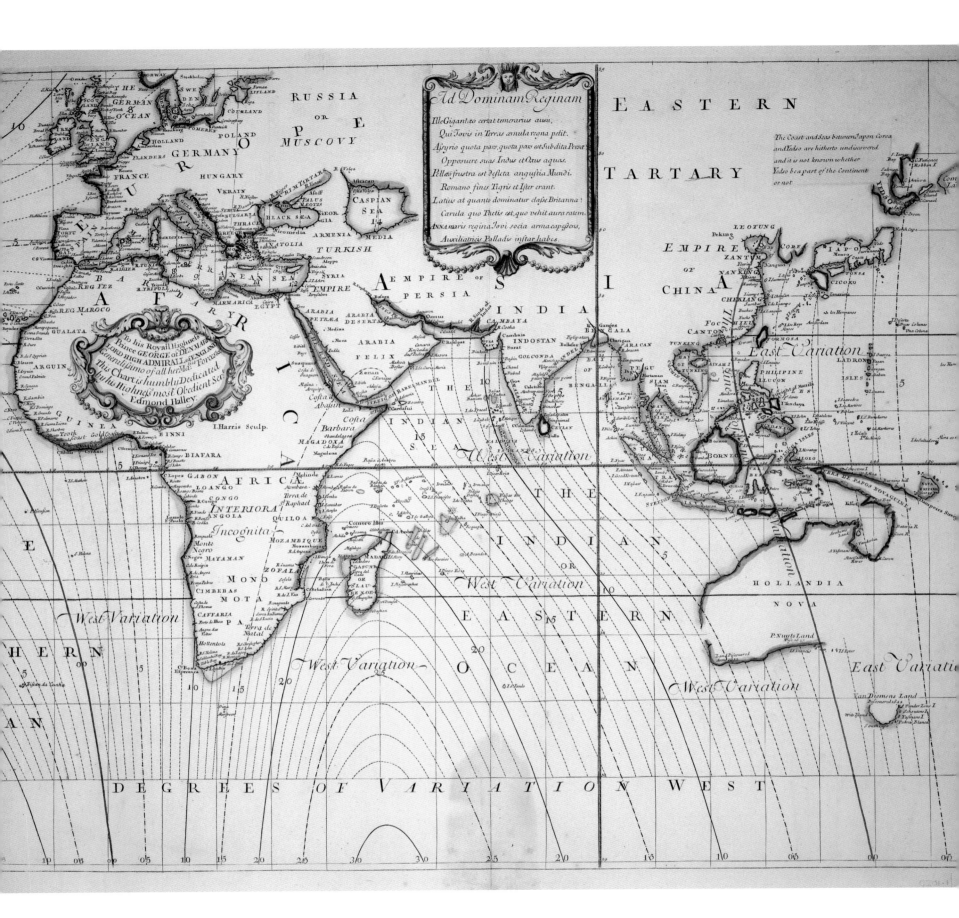

Fig. 9 – Galileo Galilei, attributed to Francesco Apollodoro, c.1602–07

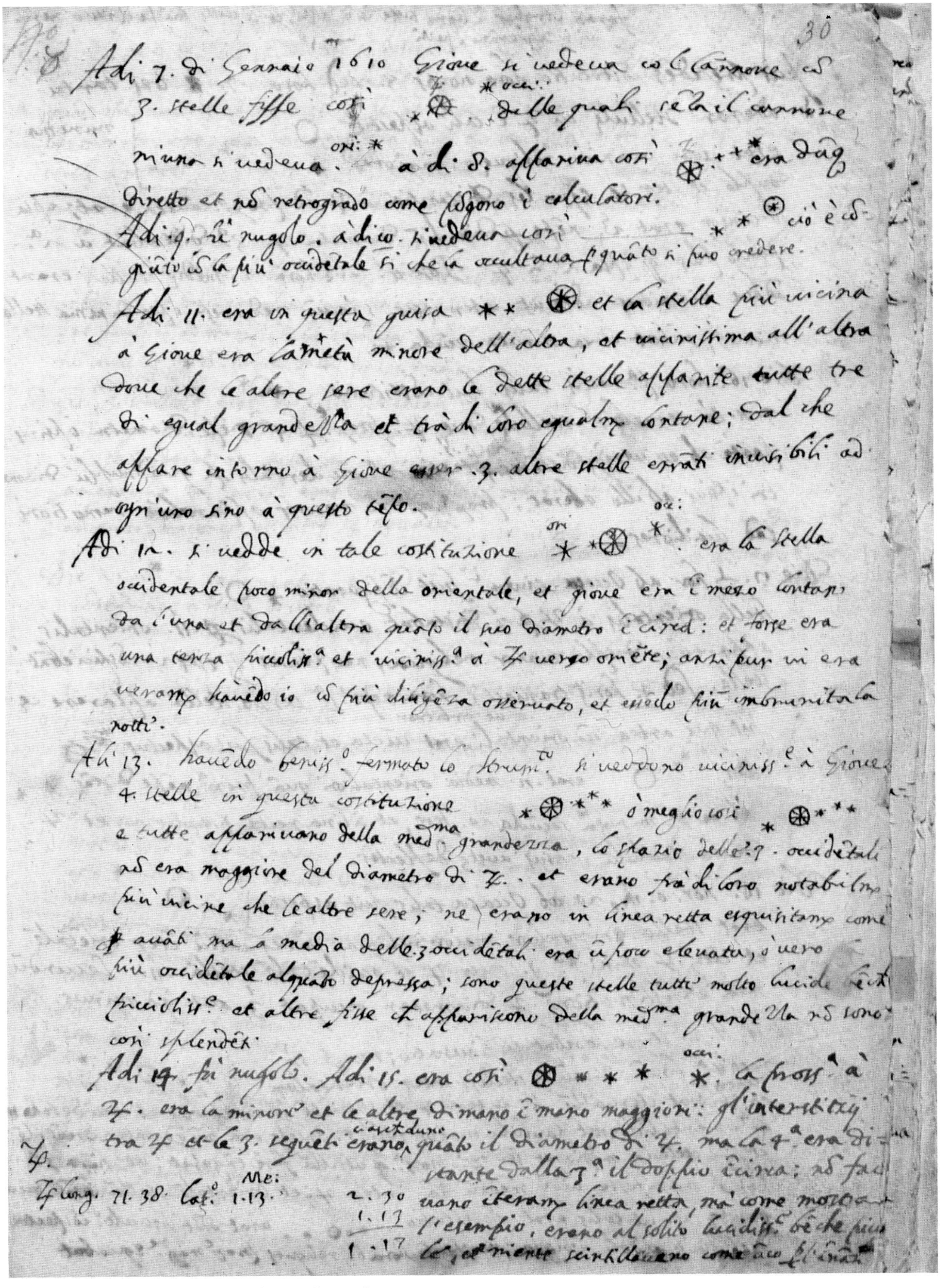

Adì 7. di Gennaio 1610 Giove si vedeva col cannone con 3. stelle fisse così … delle quali senza il cannone niuna si vedeva. à dì 8. appariva così … era dunque diretto et non retrogrado come vogliono i calculatori.

Adì 9. si nugolò. à dì 10. si vedeva così … ciò è congiunto con la più occidentale sì che la occultava per quanto si può credere.

Adì 11. era in questa guisa … et la stella più vicina à Giove era la metà minore dell'altra, et vicinissima all'altra dove che le altre sere erano le dette stelle apparite tutte tre di egual grandezza et trà di loro egualmente lontane; dal che appare intorno à Giove esser 3. altre stelle erranti invisibili ad ogni uno sino à questo tempo.

Adì 12. si vedde in tale costituzione … era la stella occidentale poco minor della orientale, et Giove era in mezzo lontano da l'una et dall'altra quanto il suo diametro in circa: et forse era una terza piccolissima et vicinissima à ♃ verso oriente; anzi pur vi era veramente havendo io con più diligenza osservato, et essendo più imbrunita la notte.

Adì 13. havendo benissimo fermato lo strumento si veddono vicinissime à Giove 4. stelle in questa costituzione … o meglio così … e tutte apparivano della medesima grandezza, lo spazio delle 3. occidentali non era maggiore del diametro di ♃. et erano trà di loro notabilmente più vicine che le altre sere; ne erano in linea retta esquisitamente come per avanti ma la media delle 3. occidentali era un poco elevata, o vero più occidentale alquanto depressa; sono queste stelle tutte molto lucide benchè piccolissime et altre fisse che appariscono della medesima grandezza non sono così splendenti.

Adì 14. fu nugolo. Adì 15. era così … la prossima à ♃ era la minore et le altre di mano in mano maggiori: gl'interstizii tra ♃ et le 3. seguenti erano ciascheduno quanto il diametro di ♃. ma la 4.a era distante dalla 3.a il doppio in circa; non facevano esquisitamente linea retta, ma come mostra l'esempio, erano al solito lucidissime benchè piccole, et niente scintillavano come anco gl'innanzi.

♃ long. 71.38 lat. 1.13 Me: 2.30 1.13 1.17

Fig. 10 – Galileo's journal of the observations of Jupiter and its satellites, 1610 which supported a telescope that could be adjusted continually to counteract the ship's movement. At least one was made and tried on board a ship in the harbour at Livorno. It impressed a member of the powerful Medici family, who apparently 'judged this invention more important than the discovery of the telescope itself'.[10] Another idea was a hemispherical vessel in which the observer could sit and which would, in theory, be kept level by floating in a bath of oil (see Chapter 3, Fig. 29). Given the small number of telescopes at this time that could show Jupiter's satellites at all, let alone a sharp image, such adaptations for use at sea were perhaps premature. Nevertheless, chairs, platforms and other devices that would ease shipboard observations continued to be explored.

Ongoing attempts to make the method workable at sea were encouraged by the successful use of Jupiter's satellites to establish longitude on land. This method began to flourish with the availability of improved telescopes and the publication of more accurate tables by Giovanni Cassini (1625–1712) in 1668. As director of the newly established observatory in Paris (Fig. 11), Cassini promoted the use of his tables on expeditions and in the mapping of France. In 1693, the Académie des Sciences published a map that compared the position of France's coastlines on the new maps with the old (Fig. 12). Although Louis XIV, it is said, complained that the astronomers had taken more territory from him than his enemies, he and the Académie continued to finance the method and ambitious expeditions to map the nation and her empire.

Cassini's method and tables were taken up elsewhere, including Britain. There, the new observatory at Greenwich focused on longitude, and its first director, John Flamsteed, produced his own tables of Jupiter's satellites, published by the Royal Society in 1683. He doubted their use could be made practical at sea but encouraged sailors to learn the method for use on coastal surveys (or, rather, berated them for having not already begun to do so). By the early eighteenth century, it was clear that the use of simultaneous observations of Jupiter's satellites to establish longitude on land could, with the best equipment and

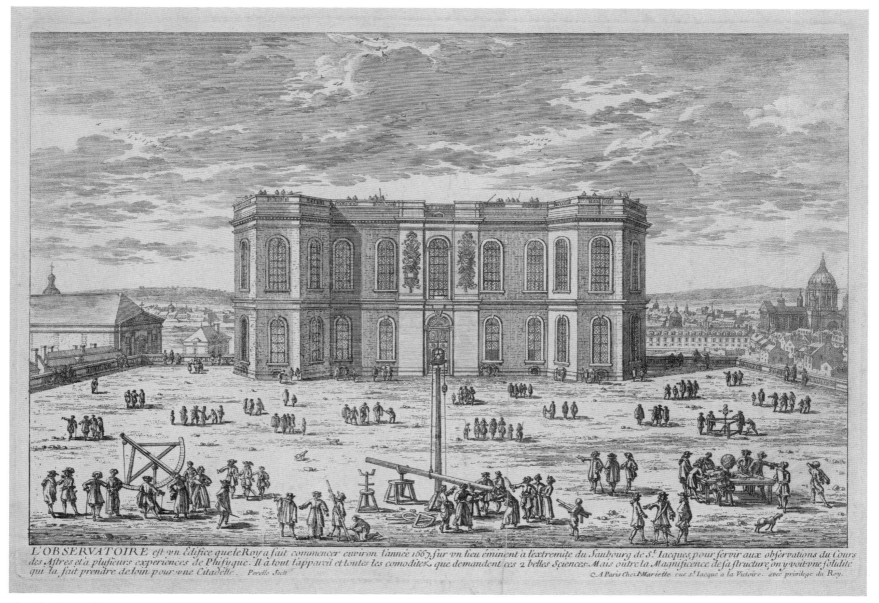

Fig. 11 – Paris Observatory, 1729. An astronomical quadrant with a telescopic sight and a large telescope with a mast and pulley for raising it are shown

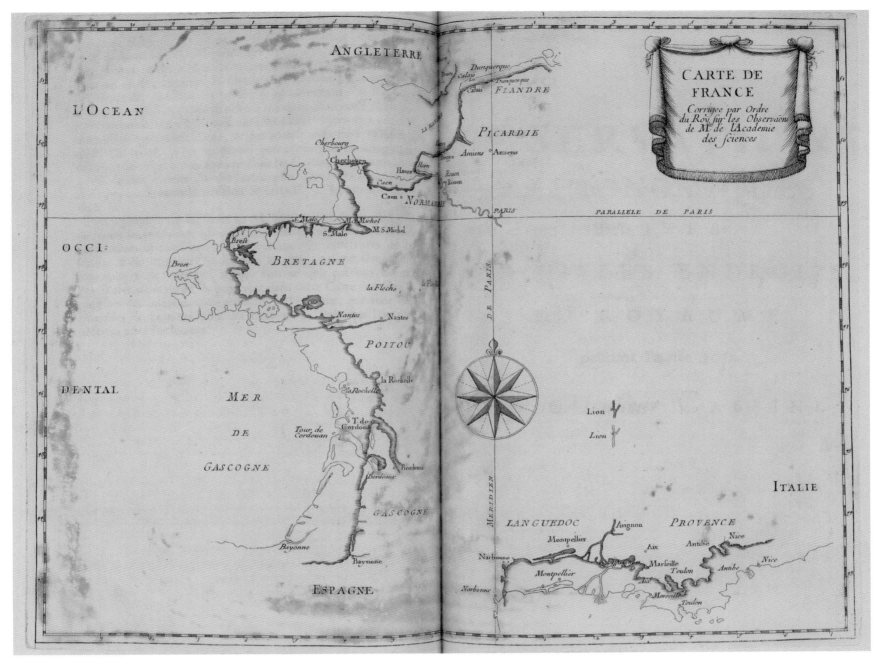

Fig. 12 – Map of France that compares the position of its coastlines on maps using new astronomical data with older maps, from *Recueil d'Observations* (Paris, 1693)

careful observers, be extremely effective. The main focus of research into astronomical methods for use at sea, meanwhile, moved elsewhere.

'the Place of the Moon'

While the leading astronomers at Paris and Greenwich had championed the use of Jupiter's satellites, the founding of both observatories had been prompted by interest in what was known as the lunar-distance method. In France it was raised by a physician and Professor of Mathematics at the Collège Royal, Jean-Baptiste Morin, in 1634, and in London in 1674 by Le Sieur de St Pierre, another Frenchman, about whom almost nothing is known beyond his title. The method had advantages over Jupiter's satellites in terms of what navigators would be required to observe at sea but it also had significant disadvantages with regard to the complexity of the Moon's motions.

The method was already more than a century old, the first description having been published by Johann Werner of Nuremburg in 1514 and clarified in Peter Apian's *Cosmographica* (1524) and *Introductio Geographica* (1533, Fig. 13), influential works that went through several editions. It made use of the well-known cross-staff (Fig. 14) to measure the Moon's position as it moved against the background of stars. The crucial measurement – the lunar distance or 'lunar' – was the angle formed between the Moon and a star. With that, plus their altitudes, an accurate reference for the positions of bright stars distributed around the night sky and an almanac predicting positions of the Moon, a navigator could find the time at the place on which the tables' data was based and subtract this from observed local time. This worked in theory but neither tables nor instruments were yet sufficiently accurate in practice, and the calculations required were actually more complex than those mentioned by Werner.

INTRODVCTIO

GEOGRAPHICA PETRI APIANI IN DOCTISSIMAS VERneri Annotationes, cõtinens plenum intellectum & iudicium omnis operationis, quæ per sinus & chordas in Geographia confici potest, adiuncto Radio astronomico cum quadrante nouo Meteoroscopii loco longe vtilissimo.

HVIC ACCEDIT Translatio noua primi libri Geographiæ CL. Ptolemæi, Translationi adiuncta sunt argumenta & paraphrases singulorũ capitum: libellus quóq; de quatuor terrarum orbis in plano figurationibꝰ Authore Vernero.

LOCVS etiam pulcherrimus desumptus ex fine septimi libri eiusdem Geographiæ Claudii Ptolemæi de plana terrarum orbis descriptione iam olim & à veteribꝰ instituta Geographis, vnà cum opusculo Amirucii Constantinopolitani de iis, quæ Geographiæ debent adesse.

ADIVNCTA est & epistola IOANNIS de Regiomonte ad Reuerendissimum patrem & Dominum D. Bessarionem Cardinalem Nicenum, atq; patriarcham Constantinopolitanum, de compositione & vsu cuiusdam Meteoroscopii armillaris, Cui recens iam opera PETRI APIANI accessit Torquetum instrumentũ pulcherrimũ sanè & vtilissimum.

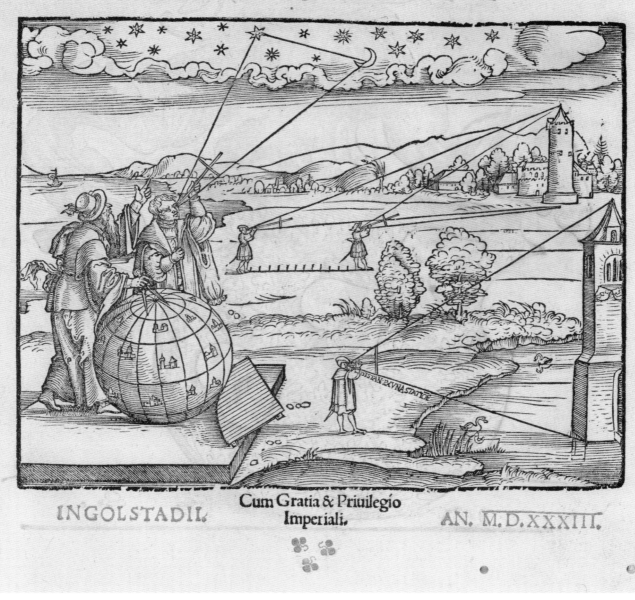

INGOLSTADII.   Cum Gratia & Priuilegio Imperiali.   AN. M.D.XXXIII.

Fig. 13 – Title page of *Introductio Geographica*, Peter Apian (Ingolstadt, 1533)

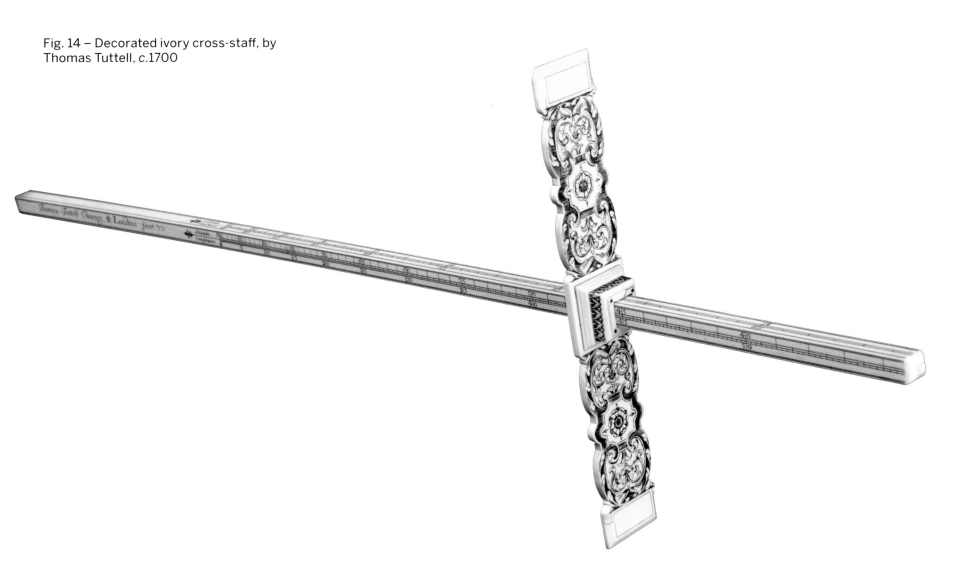

Fig. 14 – Decorated ivory cross-staff, by Thomas Tuttell, c.1700

There were some surprisingly early attempts to determine longitude by versions of this method. In July 1612, the English Arctic explorer William Baffin observed the Moon's transit, or passing, across his local meridian to determine longitude while in Greenland, but found it 'somewhat difficult and troublesome'.[11] He took the observation from land, recognizing that it would be impossible from a moving ship. Three years later, on an expedition in search of the North-West Passage, he made observations from the ship to determine longitude from the angular distance of the Moon from the Sun. The range of observations Baffin made on these voyages suggests that he was quite unusual among English mariners. For most, these forays into the complexities of astronomical navigation would have been entirely unfamiliar.

Even in the seventeenth century, the method was not well known beyond mathematical circles. Thus, when Charles II heard of St Pierre's claims to have solved the problem of longitude by using this method, a commission was appointed to examine their validity. The commissioners foreshadowed those appointed forty years later by the 1714 Act. They included the President of the Royal Society, the King's Master of Mechanics, professors of astronomy and mathematics, including Christopher Wren (1632–1723) and Robert Hooke, and other Fellows of the Royal Society. Most of them had also been responsible for judging Henry Bond's magnetic scheme the year before and it is likely that St Pierre hoped for a reward, given the commissioners' recommendation that the King 'grant some present support for Mr Bonde'.[12]

However, as happened in France, it became clear that the method was unworkable without a vastly improved catalogue of stars and theory of the Moon's motion. In both cases, the resulting recommendation was to found an observatory and appoint astronomers. Already co-opted to the Commission was John Flamsteed, a young astronomer who had impressed several Fellows of the Royal Society and had an influential patron in Jonas Moore, Surveyor-General of the Ordnance. On 4 March 1675, Charles II signed a royal warrant that appointed Flamsteed his 'astronomical observator' and charged him 'to apply himself with the most exact Care and Diligence to the rectifying the Tables of the Motions of the Heavens, and the places of the fixed Stars, so as to find out the so much desired Longitude of Places for perfecting the art of Navigation'.[13] An observatory, designed by Wren and Hooke, was built at Greenwich (Fig. 15) and Flamsteed began his long series of observations there on 16 September 1676.

It was in the second of Flamsteed's tasks that he had greatest success, although it was over a lifetime of observation and was the cause of some tribulation. His great legacy was a much larger and more accurate catalogue of 'fixed stars' than previously existed. On the way to producing this masterwork, Flamsteed published tables of Jupiter's satellites and other

PROPOSITIO XXV. PROBLEMA VI.

Invenire vires Solis ad perturbandos motus Lunæ.

Deſignet *S* Solem, *T* Terram, *P* Lunam, *PADB* orbem Lunæ. In *SP* capiatur *SK* æqualis *ST*; ſitque *SL* ad *SK* in duplicata ratione *SK* ad *SP*, & ipſi *PT* agatur parallela *LM*; & ſi gravitas acceleratrix Terræ in Solem exponatur per diſtantiam *ST* vel *SK*, erit *SL* gravitas acceleratrix Lunæ in Solem.

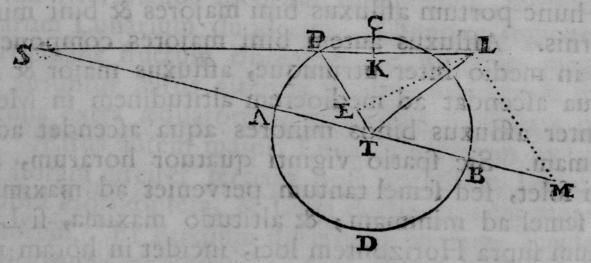

Fig. 15 (previous page) – Royal Observatory from Crooms Hill, British School, *c.*1696

Fig. 16 – The effects of the Sun on the Moon's motion, from Isaac Newton's *Philosophiae Naturalis Principia Mathematica* (Cambridge, 1713) (detail)

observations, calculations and commentaries. He also shared data with astronomers and mathematicians across Europe as part of a reciprocal, scholarly correspondence. His relationship with one of the most important figures of the period, Isaac Newton, broke down when he felt this scholarly etiquette was ignored. Newton, desperate to have access to as many observations as he could for the improvement of his theoretical work, demanded, in Flamsteed's view, too much and gave too little in return.

A serious quarrel took place over Flamsteed's catalogue. Newton, as President of the Royal Society, urged its publication and Prince George of Denmark, Queen Anne's consort and Lord High Admiral, agreed to pay. Newton, Halley and other Royal Society 'referees' for the publication gained the upper hand by persuading Queen Anne to appoint a Royal Society committee as a Board of Visitors to the Royal Observatory, with power to tell the Astronomer Royal what to observe and publish. In 1712, an edition of Flamsteed's valuable catalogue was published prematurely, without his knowledge and to his lasting fury. The process of assembling his own edition took the rest of his life and, thanks to the dedication of his wife and two assistants, appeared posthumously in 1725.

Newton's interest in Flamsteed's observations had been particularly intense when he was struggling with the second edition of his *Principia*. The first edition, with its mathematically expressed laws, including the inverse square law of universal gravitation, provided a new and essential framework for predicting the motions of the planets. However, the complexity of the Moon's motion, caused by the interplay of the gravitational influences of Earth, Sun and Moon, was great enough to defeat Newton. He struggled again with lunar theory and the so-called three-body problem for the book's 1713 edition (Fig. 16), building on observational data from Flamsteed and others and recalling later that 'his head never ached but with his studies on the moon'.[14]

Newton wished to devise a theory, based on both mathematics and empirical observations, that was accurate to two minutes of arc (that is, to one-thirtieth of a degree). He felt that this high level of accuracy was necessary in the theory in order to achieve accuracy to within one degree in practical navigation observations. However, in this he failed. Despite his best efforts, his evidence to the parliamentary committee of 1714 had to state that the Moon's 'Theory is not yet exact enough' to determine longitude at sea within one degree. Nevertheless, he gave the impression that this improvement would be forthcoming and that it was here that the long-awaited solution would lie.

'a Watch to keep Time exactly'

The accuracy of Flamsteed's observations depended on a recent revolution in timekeeping. One of the coordinates that indicates the position of stars is expressed as time: the moment at which a heavenly body passes, or transits, the observer's local meridian (line of longitude). In order to record this accurately and precisely good clocks are required. Clocks had become capable of acting as scientific instruments, known as astronomical regulators, once they incorporated pendulums, the timekeeping properties of which had been observed by Galileo. It was left to the Dutch mathematician and astronomer Christiaan Huygens (1629–95) to apply this to a clock in 1657.

The Octagon Room of the Royal Observatory at Greenwich was designed around the pendulum clocks installed there (Fig. 17). Made by Thomas Tompion, London's leading clockmaker, they were an experimental design with thirteen-foot pendulums suspended behind the room's panelling and above the dials. They were accurate enough to help Flamsteed in his first task as Astronomer Royal: to demonstrate that the Earth itself is a regular timekeeper, a prerequisite of the positional astronomy he was appointed to improve.

The timekeeper method of finding longitude at sea – a shipboard clock that would keep the time at a known location throughout a sea voyage for comparison with observations of local time – was, as Whiston and Ditton said, 'the easiest to understand and practice'. The huge strides made in the accuracy of pendulum

Fig. 17 – The Octagon Room at the Royal Observatory, Greenwich, by Francis Place, c.1676

clocks were encouraging but the technical challenges facing their application at sea were huge. Watches, less influenced by the motion of the ship, were much too inexact. As Whiston and Ditton's 1714 pamphlet put it:

> Watches are so influenc'd by heat and cold, moisture and drought; and their small Springs, Wheels, and Pevets are so incapable of that degree of exactness, which is here requir'd, that we believe all wise Men give up their Hopes from them in this Matter. Clocks, govern'd by long Pendulum's, go much truer: But then the difference of Gravity in different Latitudes, the lengthening of the Pendulum-rod by heat, and shortening it by cold; together with the different moisture of the Air, and the tossings of the Ship, all put together, are circumstances so unpromising, that we believe Wise Men are almost out of hope of Success from this Method also.[15]

It had 'been so long in vain attempted at Sea, that we see little Hopes of its great usefulness there'.[16] Dependence on a single clock was also potentially dangerous if there were no means to check its performance.

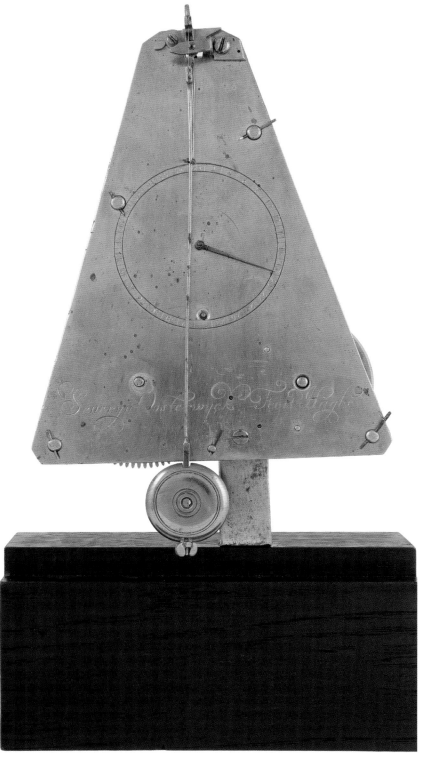
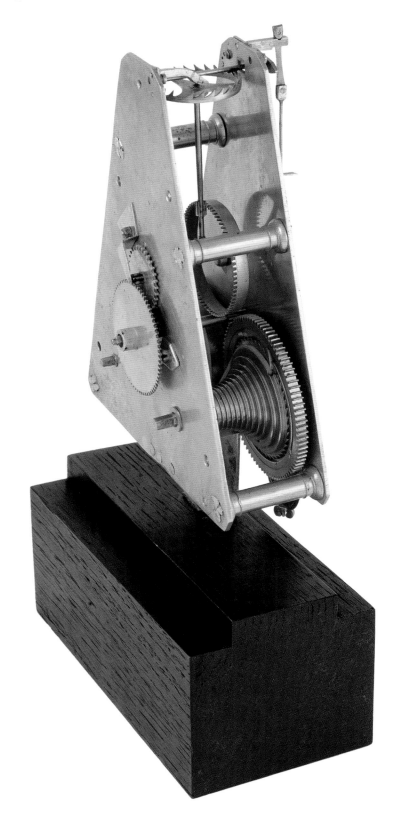

Fig. 18 – Marine timekeeper, by Severyn Oosterwijck, *c*.1662 based on the designs of Alexander Bruce and Christiaan Huygens

who, in the 1670s, were involved with the invention of the balance spring, crucial to the development of accurate watches. There are hints that Huygens saw this as a potential alternative for a timekeeper solution to the longitude problem but, because springs were too greatly affected by changes in temperature, it was not one he pursued.

Huygens died in 1695, having made some huge practical and theoretical advances but without a clock having yet been taken up as a usable tool at sea. His influence on those who followed, through his publications, manuscripts or collaborators, was enormous. One follower was Lothar Zumbach de Koesfelt, a Dutch physician, mathematician and musician, who described a sea-clock in 1714. It was later improved by his son Conrad, who in 1749 also designed a clock that used a glass container to control its temperature (Fig. 21). Another was Henry Sully, who, trained in England and working on the Continent, experimented with a marine clock and watch (Fig. 22). Sully made ample use of both French and English networks, not least the Royal Society's chief authority on instruments and timekeepers, George Graham (1673–1751). While Sully's clock had mechanisms that made it portable and minimized the effects of temperature and gravity, Graham's report and trials were to show that a ship's motion in open sea would fatally influence its pendulum.

see page 43, Fig. 6

In 1714, then, there were several potential solutions to the problem of finding longitude at sea. Most of them were theoretically viable and had already been researched by mathematicians, astronomers, artisans and mariners. In many ways the ground had been prepared: the next, crucial step may not have seemed all that distant. However, all of the methods presented practical problems, most particularly due to the conditions under which observers or instruments had to perform at sea. It was a practical and technical set of problems, and the government hoped that these could be met so that the methods 'true in theory' might be made 'Practicable and Useful at Sea'.

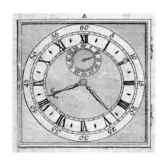

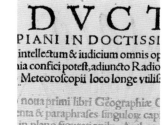
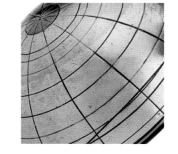

While much previous work had taken place in Spain, Holland, Venice and elsewhere, the momentum had decisively shifted towards Britain. Ideas and skills were available, particularly in London with its flourishing instrument trade, maritime interests, places of open discussion, such as coffee-houses, and widespread access to print. These opportunities were seized by many of those with ideas, mechanisms and projects relating to longitude. The Royal Society, Astronomer Royal and Commissioners of Longitude were marked out as the key figures from whom to seek interest, approval and, perhaps, a route towards one of the 1714 Act's enormous rewards.

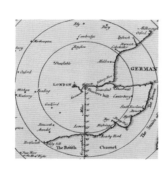

CHAPTER 3

ON TRIAL

I understand not Mathematicks, but have been formerly troubled too much with Projectors of the Longitude to my great Mortification and some Charges by encouraging them ... One of my Projectors cut his Throat, and the other was found an Imposter.

Jonathan Swift, letter to John Wheldon, 27 September 1727[1]

The likely effect of the Longitude Act must have been uncertain as it went through Parliament, but within weeks it was clear that it had caught the public's attention. Schemes of all sorts came under public scrutiny – a mixed bag of ideas, some more plausible than others and many held up for ridicule. In an age awash with projects of all sorts, getting these ideas taken seriously and gaining support required tenacity and robust strategies.

It would be two decades before a genuinely promising scheme emerged in the work of John Harrison (1693–1776). From the mid-1730s, his sea-clocks began to garner praise and the support of the Commissioners of Longitude. Even so, it took him another twenty years to complete a watch that was ready for a decisive trial. Over the same period, two other contenders emerged in Tobias Mayer's work on the Moon and Christopher Irwin's marine chair for viewing Jupiter's satellites. The three methods would finally go head-to-head in 1763–64 in a sea trial to Barbados, which established marine timekeepers and the measurement of lunar distances (or 'lunars') as the methods to back.

   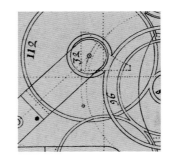 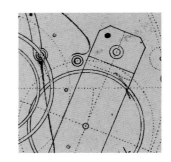
  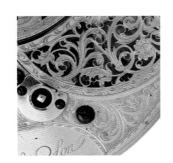  

> *Had we of Archimedes's Lumber,*
> *Enough to make a Chair for Slumber,*
> *We'd find by Lines in Cucumber*
> *Longitude.*
>
> *A Hymn to the Chair* (1732)[2]

  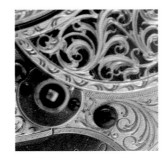  

An ABSTRACT of the Act of Parliament concerning the Discovery of the
LONGITUDE,
And other Improvements of NAVIGATION.

ENACTED, That the Lord High Admiral of *Great Britain*, or the first Commissioner of the Admiralty; The Speaker of the House of Commons; The first Commissioner of the Navy; The first Commissioner of Trade; The Admirals of the Red, White, and Blue Squadrons; The Master of the Trinity-House; The President of the Royal Society; The Royal Astronomer of *Greenwich*; The Savilian, Lucasian, and Plumian Professors of the Mathematicks in *Oxford* and *Cambridge*; All these (fourteen) for the time being;

 THOMAS Earl of *Pembroke* and *Montgomery*;
* PHILIP Lord Bishop of *Hereford*;
* GEORGE Lord Bishop of *Bristol*;
 THOMAS Lord *Trevor*;
 Sir THOMAS HANMER Bart.
* FRANCIS ROBERTS Esq;
* JAMES STANHOPE Esq;
* WILLIAM CLAYTON Esq;
* WILLIAM LOWNDES Esq; or

Any five of those Commissioners may receive Proposals; and, if satisfied of the probability of a Discovery of the Longitude, may certify the same to the Commissioners of the Navy with the Author's Names; On producing which Certificate the Navy Commissioners shall make Bills for any Sum not exceeding two thousand Pounds for making Experiments, payable by the Treasurer of the Navy who is immediately to pay the same. And after Experiments made, the Commissioners named in the Act shall determine how far the same is found practicable, and to what degree of Exactness. Ten thousand Pounds to the Person who determines the Longitude to one Degree of a great Circle or sixty Geographical Miles; Fifteen thousand Pounds if within two Thirds of that Distance; And twenty thousand Pounds if within half of the said Distance. One half of such Reward to be paid when the major part of the Commissioners agree that any such Method extends to the Security of Ships within eighty Geographical Miles of the Shores, which are Places of greatest Danger; And the other half when a Ship (by the Commissioners Appointment) shall sail from *Great Britain* to the *West-Indies*, to such Port as shall be named, without losing the Longitude beyond the Limits mentioned. As soon as this Experiment shall be made by this Ship so sent out, and found practicable; on the Commissioners Certificate thereof the remainder of the Money to be paid by the Treasurer of the Navy. And though any Proposal shall not on Trial be found of so great use; Yet if it be found of considerable Use to the Publick, The Author to have such Reward as the Commissioners shall think reasonable; to be paid by the Treasurer of the Navy, on such Certificate as aforesaid.

FINIS.

Fig. 1 – A summary of the 1714 Longitude Act circulated as a flyer

Projectors in public

Within days of royal assent being granted, details of the Longitude Act appeared in newspapers and periodicals, with abridged versions circulating as flyers (Fig. 1). The public response was just as rapid. Some longitude schemes had been published before, but the number increased dramatically and older proposals were quickly republished. Generally they repeated the known contenders: magnetic variation; marine timekeepers; improved methods of dead reckoning; and astronomy. However, as the newly appointed Commissioners of Longitude knew, the basic theories were largely sound. It was detailed solutions to the practical problems of being on a ship that they were looking for.

For the authors of longitude schemes, gaining credibility was the challenge, with ideas of all sorts competing for investors. Every day, Daniel Defoe noted, projectors were coming up with 'new Contrivances, Engines, and Projects to get Money'.[3] Some were honest, he conceded, but potential backers were understandably cautious, particularly after the notorious financial disaster of the South Sea Bubble of 1720, said to have ruined people from all walks of life. Such wariness extended to longitude schemes. Looking back on the first ten years since the Longitude Act, one mathematical author claimed that the rewards

> had no other effect than setting all heads and hands at work, and producing, for the sake of interest, abundance of quacks and impossible schemes ... in the year 1722 the discovery of the longitude, was made one of the bubbles in *Change-Ally*, and put up to be sold to the best bidder. Many brokers, crazy projectors, and poor crafty knaves, cryed it at *Garraway's Coffee-House*, along with ... *South-sea stock*.[4]

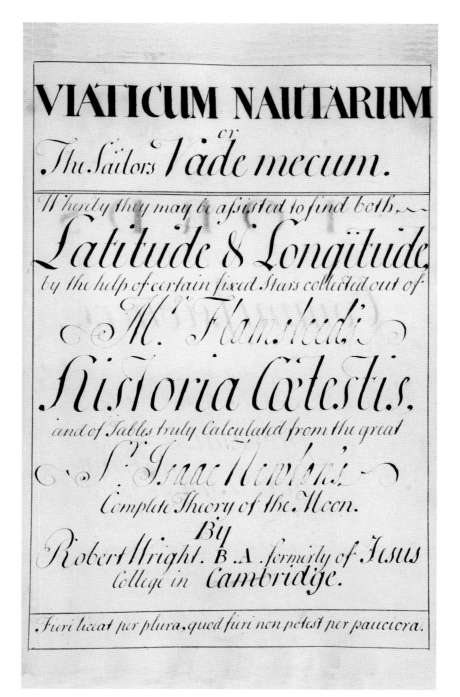
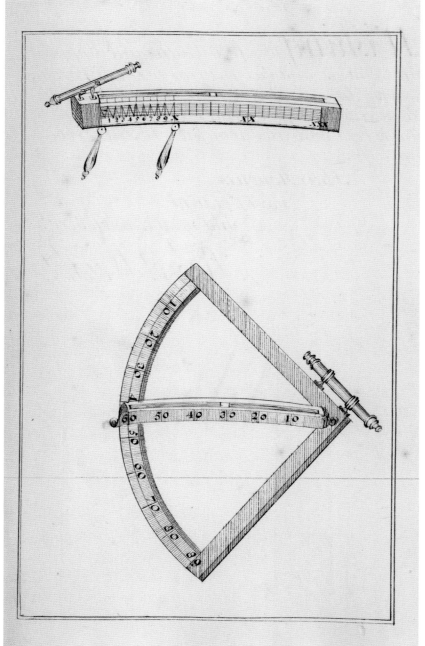

Fig. 2 – 'Viaticum Nautarum or The Sailor's Vade Mecum', by Robert Wright, 1726, a longitude scheme that was sent to Isaac Newton

Anyone with pretensions to the rewards needed to impress the Commissioners or some other influential patron. Some projectors took the bull by the horns and approached individual Commissioners directly, sending schemes (Fig. 2) or visiting in person. As Astronomer Royal and an *ex officio* Commissioner, John Flamsteed received more than his fair share: people knew where to find him and his Greenwich appointment made him a longitude expert. In August 1714, two Derbyshire preachers came to the Observatory in Greenwich Park with a scheme to use the refilling of evacuated vessels to measure time accurately. Flamsteed was polite, but 'refused to give them my hand to testifie that I had seen their proposealls and advised them not to print them'. Another hopeful inventor, Case Billingsley, outlined a platform for shipboard observations; the astronomer carefully explained its flaws, while Billingsley 'held his peace which is as much as I can expect from persons that have swelled themselves with the hopes of getting twenty thousand pounds'.[5]

Publishing was the main way of promoting an idea and, hopefully, making money, with the printed word lending a certain gravitas. Within three months, a 'Swarme of *hopefull Authors*' had proposals rolling off the press, many of them directly addressed to the Commissioners. They included the Venetian mathematician Dorotheo Alimari, whose observing instrument Flamsteed considered 'one of the worst contrivances for taking the height of the Sun or Stars that ever was thought of' (Fig. 3).[6]

Isaac Newton, the country's leading man of science and a Commissioner by virtue of his presidency of the Royal Society, was another obvious target and was soon weighed down with schemes. He resolutely insisted that astronomy held the solution and that any other method was at best subservient. As he wrote to one thick-skinned enquirer about timekeepers:

> I have told you oftner than once that it is not to be found by Clock-work alone ... Nothing but Astronomy is sufficient

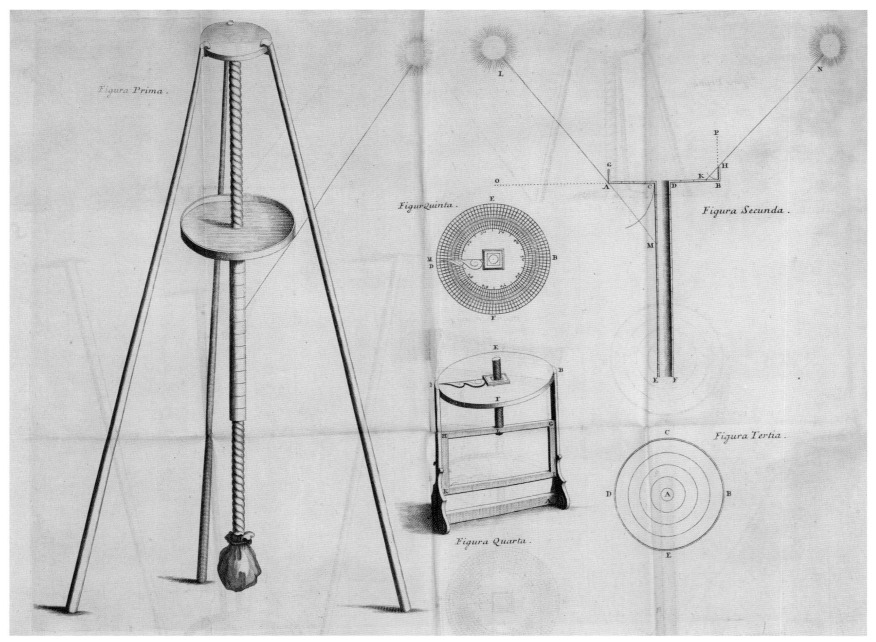

Fig. 3 – Dorotheo Alimari's observing instrument, from *The New Method Proposed by Dorotheo Alimari to Discover the Longitude*, by Sebastiano Ricci (London, c.1714)

Fig. 4 – A satirical print, *The Coffee House Politicians*, c.1733, poking fun at some of the goings on in London's coffee-houses

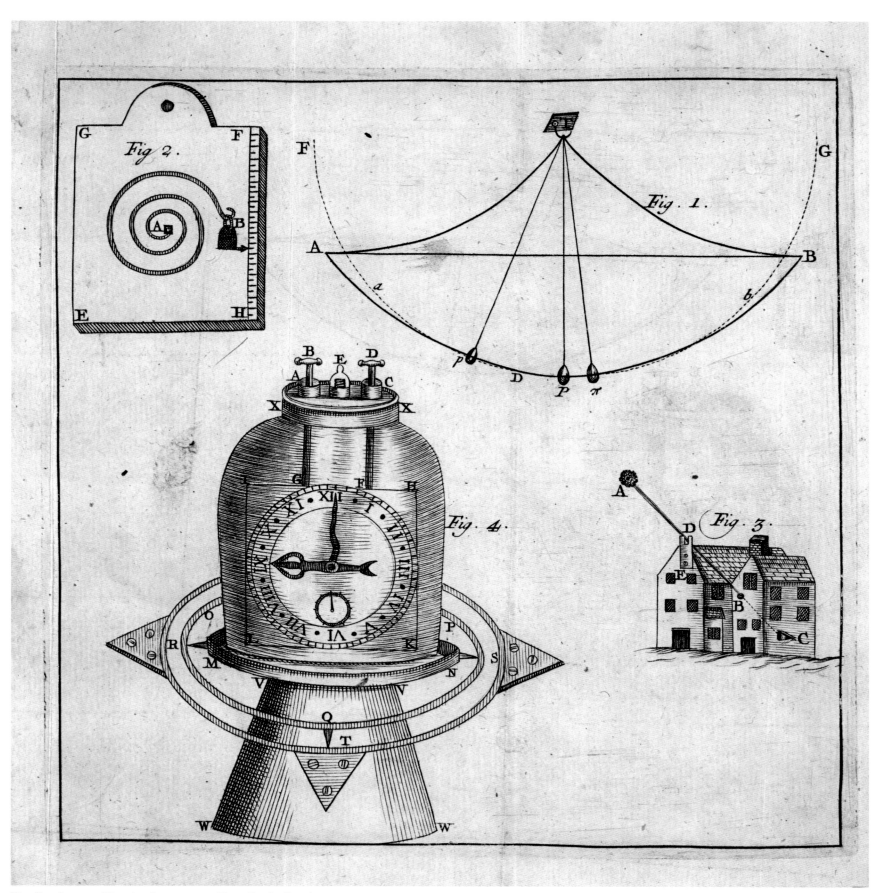

Fig. 5 – Jeremy Thacker's proposed longitude timekeeper, from *The Longitudes Examin'd* (London, 1714)

for this purpose. But if you are unwilling to meddle with Astronomy (the only right method & the method pointed at by Act of Parliament) I am unwilling to meddle with any other methods then the right one.[7]

While individual Commissioners did receive and respond to proposals, there was general uncertainty about the process and mounting criticism of the lack of conspicuous action. Jane Squire, the only woman known to have proposed a scheme under her own name, complained that:

If this Commission, who in near twenty Years have never thought fit to meet as Commissioners, it would be a charitable piece of Justice to inform the World of it, that this Act of Parliament may no longer remain the Destroyer of the most valuable of all Goods, Time.[8]

Squire's ideas were somewhat unusual. Hoping to help humanity reclaim the knowledge lost when the Tower of Babel was destroyed, she offered a new language as the basis of a more profound understanding of time and place. It would be taught to children using orange peel, special playing cards and building blocks, she said. Longitude, as a product of time, would be easily found as a result. Family and religious connections gave Squire a way in to the Commissioners, particularly Hans Sloane, by then President of the Royal Society, and Thomas Hanmer, a Commissioner named in the Act. Hanmer suggested she publish her ideas, emphasizing that any proposal should undergo 'the Scrutiny of all the great Professors of the Sciences of Astronomy and Navigation', as well as 'stand the Test of Practice'.[9] Squire died without reward, however.

Others explored less direct routes. The mathematician John French already had a scheme in 1706, when he persuaded Queen Anne's husband, Prince George, to test a compass in which fire deflected the needle. With renewed impetus from the Act of 1714, he wrote to the Royal Society, then solicited public support in print and finally applied to the Prince of Wales (later George II) for a royal patent. The Prince replied that only the Commissioners could help him and there the matter rested. Coming to London with a different magnetic scheme, the Welsh physician Zachariah Williams failed to impress the Commissioners, so he wrote to the Duke of Chandos, who was known for investing in projects of all sorts. Chandos passed him on to the natural philosopher John Desaguliers, who was not impressed either.

Where approaches to specific Commissioners or investors failed, a public appeal was worth trying. Reports and advertisements for supposedly successful longitude methods, and for the books describing them, appeared in all the newspapers. In late July 1714, William Hobbs advertised his 'Horologe', a timepiece impervious to heat and cold. All he needed, he told readers of the *Daily Courant*, Britain's first daily newspaper, was a backer so that he could make one. By November he was offering to show the finished clock at one of London's coffee-houses. It was an idea that Flamsteed had dismissed.

William Whiston also made extensive use of newspapers and periodicals, as he had in lobbying with Humphry Ditton for the Longitude Act. In October 1714, the *Post Man* announced that every Saturday night a 'Ball of Fire' would be thrown a mile in the air from Blackheath. Whiston asked readers within sixty miles of London to note its direction and height (see Chapter 2, Fig. 4). The following year he asked them to record the time between seeing the flash and hearing the sound of the mortars. He tried again two years later but had little success in gaining responses.

Whiston had other strategies up his sleeve too. In 1715, he tried raising money by subscription for a coastal survey using his rockets. Two years later, he presented the rocket scheme to the Mayor and Court of Aldermen of the City of London, but when they consulted Edmond Halley he declared it unworkable. Whiston then published a tract on longitude by magnetic inclination in 1719, securing over £470 from subscribers including the royal family, the Duke of Chandos and Martin Folkes, a future President of the Royal Society. By 1724, he was proposing the use of solar eclipses, and, ten years later, a new reflecting telescope for observing Jupiter's satellites, which he presented at the Royal Society. Finally, when in 1739 he revived his idea for a coastal survey using magnetic variation, the Commissioners offered him £500, and patrons including the First Lord of the Admiralty gave another £175. At the time, Whiston was only the second projector to be rewarded under the Longitude Act.

Whiston was also a doyen of London's coffee-houses, which had become popular places for relaxation, chat and business. As one regular wrote, they made 'all sorts of People sociable, they improve Arts, and Merchandize, and all other Knowledge'.[10] They were places where projects of all sorts could be read about in books and newspapers (Fig. 4), discussed among the clientele or with the projectors themselves, or presented in lectures. Longitude and navigation had long been topics for coffee-house debate, with James Hodgson, a former assistant at the Royal Observatory, giving talks at Jones's in Cornhill by 1703, later moving to the Queen's Head Tavern and the Marine Coffee-house near the Royal Exchange. He even anticipated Whiston in mounting experiments with a 'large gun' at Shooter's Hill and the Royal Observatory, a method by which he believed a ship's position could be determined.[11] By the 1710s, other navigational lecturers included both Ditton and Whiston, the latter still offering talks on latitude and longitude at the Temple Exchange Coffee-house three decades later.

Some intriguing schemes emerged from this melting pot of ideas. In November 1714, advertisements appeared for Jeremy Thacker's *The Longitudes Examin'd*, which described a clock sealed in a vacuum to protect it from external influences (Fig. 5). Thacker's book was horologically knowledgeable and notable for applying the term 'chronometer' to a marine timekeeper. But perhaps it was a satire: his 'pretty machine' would 'I am (almost) sure ... do for the

Fig. 6 – The final plate, set in Bedlam, of William Hogarth's *A Rake's Progress*, 1735, showing one inmate drawing longitude schemes on the wall

Fig. 7 – Precision long-case regulator, by John Harrison, 1726

Longitude' and he confessed that, 'If it be ask'd why I wrote the Book at all, I'll frankly answer, *That I wanted Money*'.[12] It read like a sly take on projecting and made fun of other published schemes.

As Thacker suggested, many proposals were not worth the bother. Indeed, to those trying to get their ideas noticed, the plethora of poorly conceived ideas was in itself problematic. Two hopeful authors, William Ward and Caleb Smith, wrote that these 'Idle Schemes and Chimerical Projects' had

> brought so much Disgrace on the Projectors, that every *Attempt* to solve this *valuable Problem*, is now ridiculed as the effect of a weak, or a distempered Brain: The Thing itself has been so long the *Reproach of Art* ... that the greatest part of Mankind look on it as an *Impossibility*.[13]

Critics agreed: finding longitude at sea really was impossible. Jonathan Swift's Gulliver would 'see the Discovery of the *Longitude*, the *perpetual Motion*, the *Universal Medicine*, and many other great Inventions, brought to the utmost Perfection', only if he became immortal like the Struldbrugs of Luggnagg; in other words, never.[14] Projectors and investors could, therefore, expect only financial, and perhaps mental, ruin. Thus George Lyttelton's supposed *Letters from a Persian in England* described a resident of Bedlam lunatic asylum who 'had quitted Poetry, and taken to the Mathematicks, by the means of which he had found out the Longitude, and expected to obtain a great Reward'.[15] Most famously, William Hogarth linked longitude and madness in the final scene of *A Rake's Progress* (Fig. 6). His Bedlam inmates included a man sketching Whiston and Ditton's rocket proposal and other longitude ideas on the hospital wall.

Having gained notoriety for its role in the creation of the 1714 Act, Whiston and Ditton's proposal attracted particular attention. John Arbuthnot wrote to Swift that it was 'the most ridiculous thing that ever was thought on' and had spoiled an idea for their satirical creation, Martinus Scriblerus.[16] Nonetheless, Scriblerus was credited with schemes including '*Perpetuum Mobiles, Flying Engines*, and *Pacing Saddles*; the Method of discovering the *Longitude* by *Bomb-Vessels*',[17] while a scatological song credited to the same authors included the lines

> The Longitude mist on
> By wicked *Will Whiston*.
> And not better be hit on
> By good master *Ditton*.
> So *Ditton* and *Whiston*
> May both be bep-st on;
> And *Whiston* and *Ditton*
> May both be besh-t on.[18]

While they were neat rhymes, 'be-pissed on' and 'be-shit on' were hardly sophisticated, yet they suited an age obsessed with bodily parts and functions. It was a mentality for which longitude was a gift. In a poem lampooning longitude projectors, a cuckolded astronomer almost discovers his unfaithful wife with her lover and asks, 'How far from *Mars* is *Venus* off?'

> About nine Inches and a half,
> (Replies the Dame:) Sleep but one Hour,
> The LONGITUDE I shall secure.[19]

Yet, amid the satire, plausible longitude schemes did eventually emerge, some promising enough for practical trials. In 1719–20, Admiralty-sponsored tests of a 'fluid quadrant' by Jacob Rowe even led to attempts to introduce new legislation. Whiston wrote that the Commissioners met to discuss Rowe's ideas as well, although no other evidence of this intriguing possibility has come to light. Then, in the 1730s, a scheme emerged with sufficient credibility for the Commissioners to consider a reward. The game changer was a clockmaker from Lincolnshire.

John Harrison and marine timekeeping

Little is known of John Harrison's early life and education, except that he learned his father's trade of carpentry. This undoubtedly gave him some of the skills necessary for making clocks, but how and why he became interested in them is still unclear. Possibly he was inspired by contact with the clock-making trade in Hull, just across the river Humber from his home in Barrow, north Lincolnshire. What is known is that Harrison was making clocks by the time he was twenty, and that they were constructed almost entirely of wood. One of them, built for the stables at the nearby Brocklesby Park estate, was revolutionary. It ran, and still runs, without oil and had a new form of escapement, the part that feeds in the energy to keep the pendulum swinging. Harrison's design, called a grasshopper escapement – after its distinctive jumping motion – was almost frictionless.

Reducing friction was crucial for the success of Harrison's early timekeepers. The oils available at the time were the curse of clockwork, causing mechanisms to run inconsistently and break down. From his understanding of the properties of woods, however, Harrison realized that the tropical hardwood lignum vitae, which contains a natural lubricant, could be used for the bearings and would allow him to dispense with oil entirely.

Success at Brocklesby Park encouraged Harrison to design precision pendulum clocks, of which he built three in the 1720s (Fig. 7). All incorporated his friction-reducing ideas and eventually also included a device that dealt with temperature variation. For a pendulum-driven clock, the time of the pendulum swing depends on its length. If the pendulum rod is made of metal, it lengthens as the metal expands in warmer conditions, causing the clock to beat slower and lose time. Harrison got round this by using a pendulum consisting of a series of rods of two different metals, brass and steel, connected in such a way that the expansion of the metals cancelled each other out and the pendulum's effective length remained constant. This is known as a gridiron pendulum. Once adjusted, Harrison's clocks ran with unprecedented accuracy and reliability.

Harrison later wrote that it was in 1726 that he learned about the longitude rewards and turned his attention to designing a timekeeper that could work properly on a ship. This would dominate the rest of his life, with his attempts focused on the challenges that had so taxed Huygens, Sully and other clockmakers before him: reducing friction; compensating for the effects of temperature; and isochronism – ensuring that each beat of a timekeeper took the same time.

Armed with his ideas for sea-clocks, Harrison came to London in about 1727–28, looking for support and the promised rewards. As Harrison tells it, he began with Edmond Halley, Flamsteed's successor as Astronomer Royal. Halley received him warmly at Greenwich but felt unable to judge Harrison's work and so sent him to George Graham. After an unpromising start, Harrison eventually impressed the London clockmaker, probably with his elegant method for temperature compensation, to the extent that Graham even offered a loan to help him develop his ideas. From then on, Graham would be Harrison's main supporter in London, giving advice and access to books and contacts. These would be crucial as his work progressed.

For the next few years Harrison worked in Barrow on a marine timekeeper, now known as H1 (Fig. 8), probably helped by his brother James. It was an extraordinarily complex machine with over 1440 parts – over 5400 if you count each chain segment separately. In essence, it was a portable version of his pendulum clocks and, like those clocks, used a grasshopper escapement, lignum vitae bearings and a gridiron pendulum. It also had several further refinements to counteract the effects of motion at sea: it used a spring rather than a hanging weight to power the clock and the bar balances (which look like dumbbells) were connected in such a way that the effects of motion on one would be counteracted by the other.

After testing the clock on the river Humber, Harrison proudly brought it to London in 1735 and installed it in Graham's workshop to be shown to London's scientific community. Backed by a trusted member of that community, the clock and its maker quickly impressed all who came to see it: 'The sweetness of its motion', wrote the antiquarian William Stukeley, 'cannot be sufficiently admired'.[20] Graham's close contacts with the Royal Society helped as well, since within a short time the Society issued a certificate praising the clock's potential. At last, it seemed, here was a timekeeper that might be used to determine longitude at sea. A trial was called for.

In May 1736, Harrison and H1 were therefore taken aboard the *Centurion* (Fig. 9), about to set sail for Lisbon. As the First Lord of the Admiralty told her captain,

> The Instrument ... has been approved by all the Mathematicians in Town that have seen it, (and few have not) to be the Best that has been made for measuring Time; how it will succeed at Sea, you will partly be a Judge.[21]

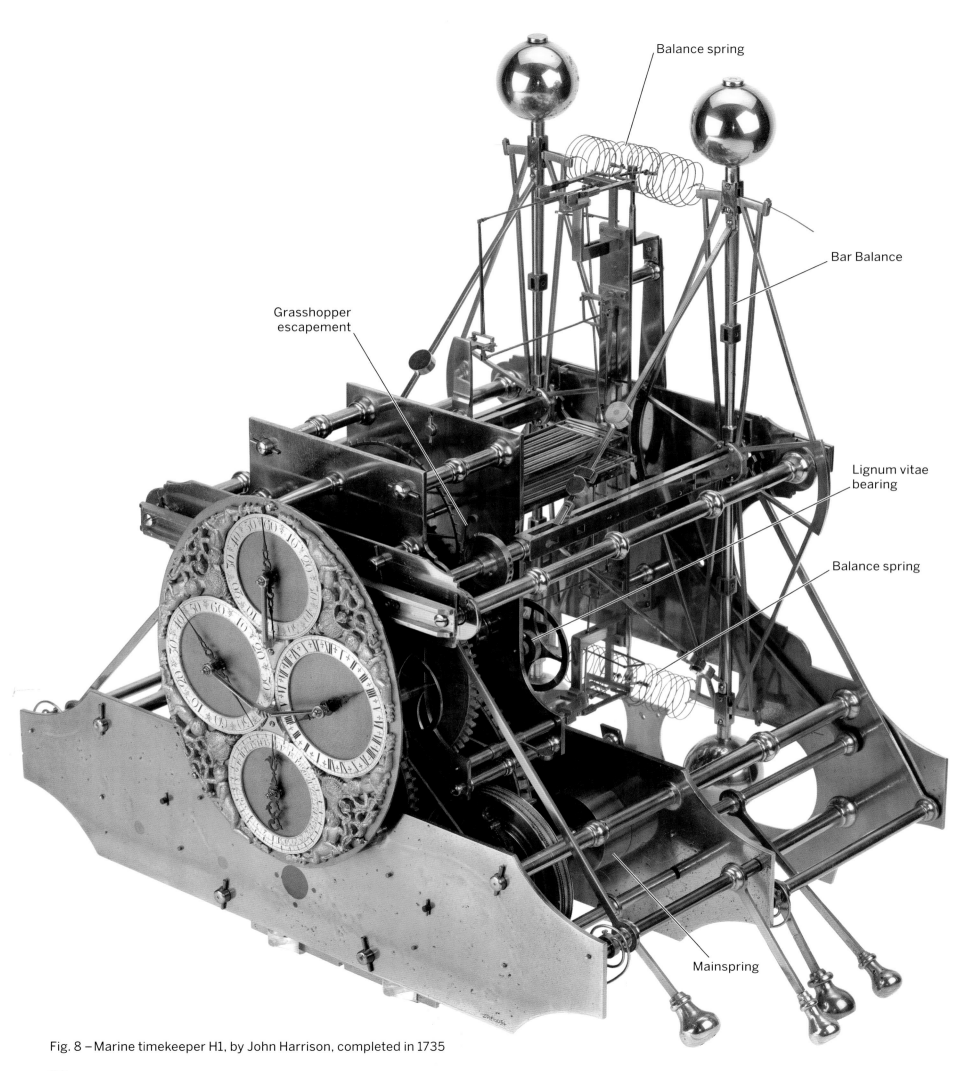

Fig. 8 – Marine timekeeper H1, by John Harrison, completed in 1735

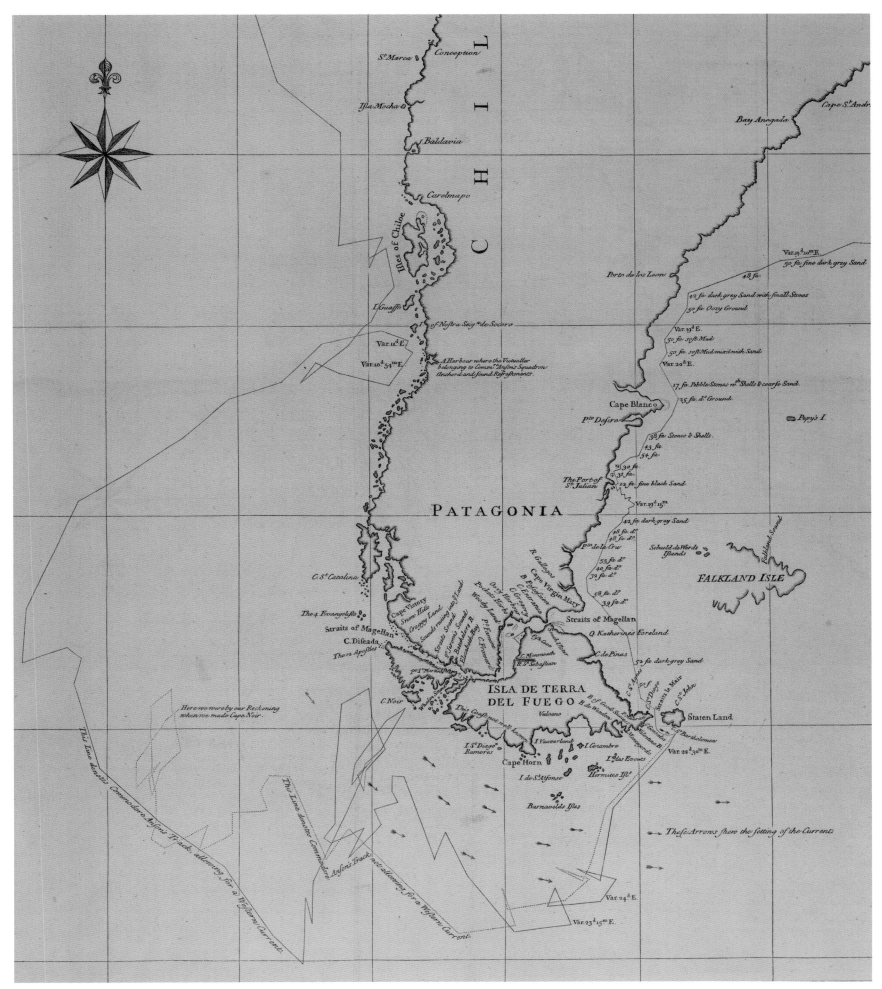

Fig. 11 – The estimated and actual tracks of the Centurion around Cape Horn in 1741, from 'A Chart of the Southern Part of South America with the Track of the Centurion', by Richard Seale, 1748 (detail)

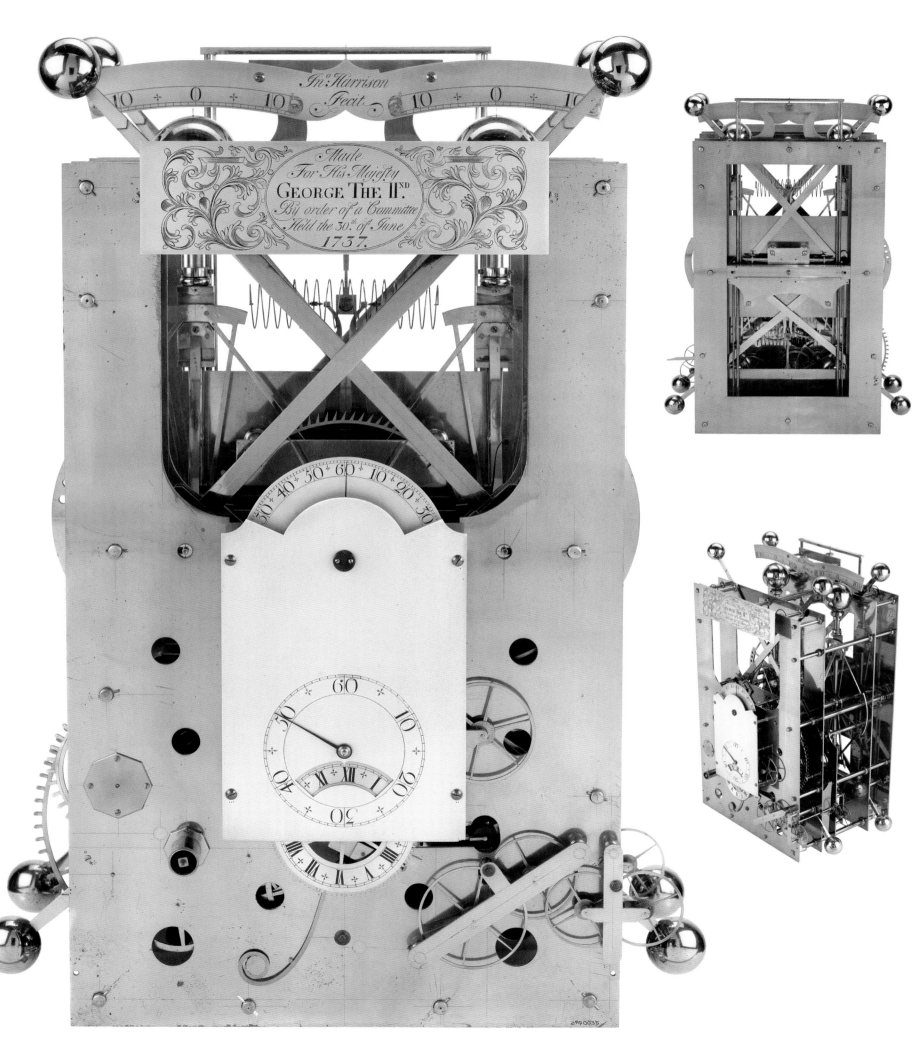

84

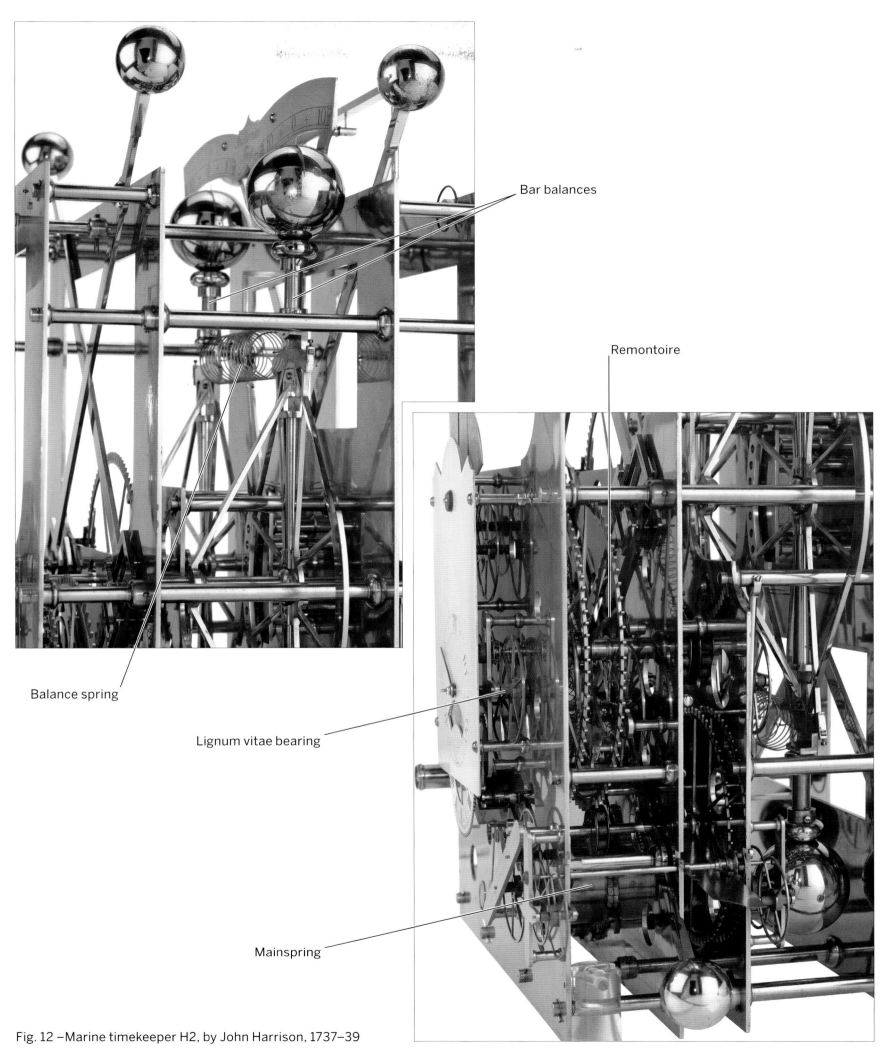

Fig. 12 – Marine timekeeper H2, by John Harrison, 1737–39

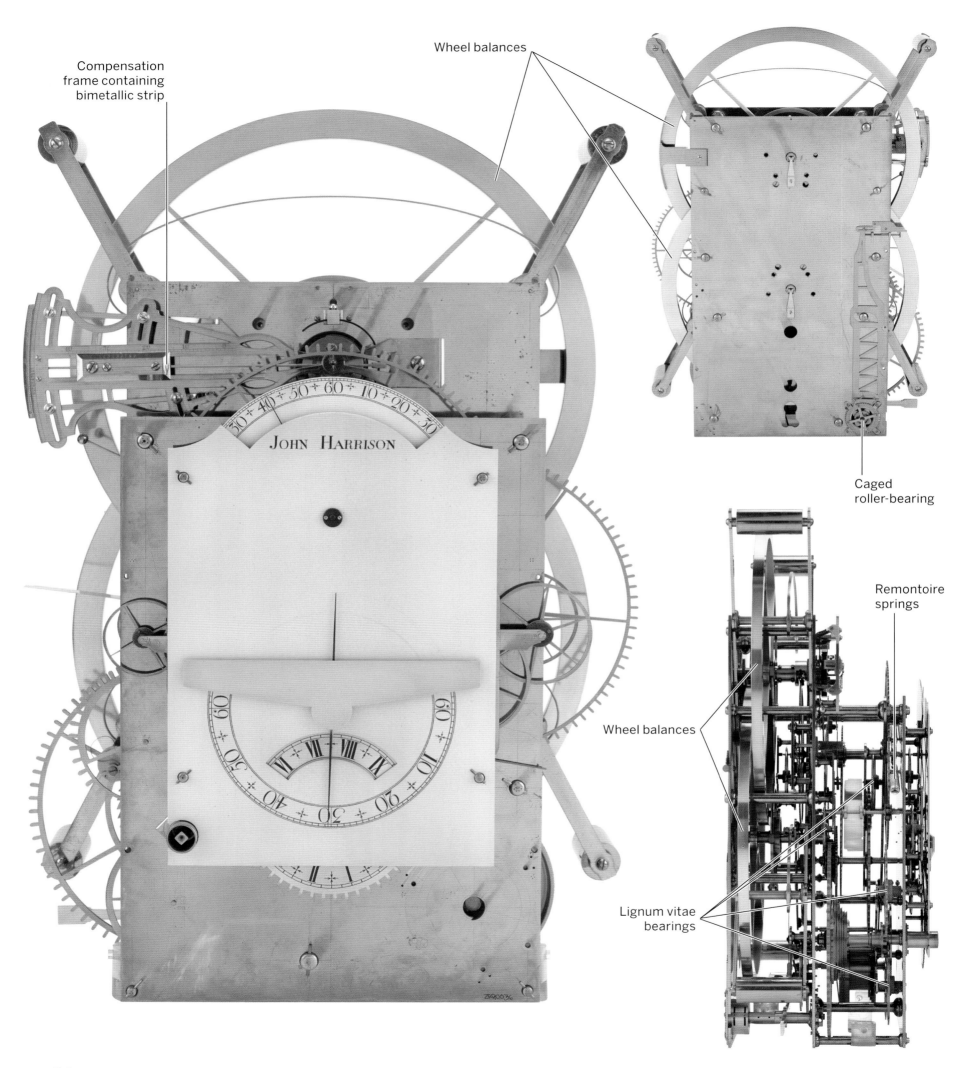

Fly of remontoire

Remontoire spring

Lignum vitae rollers

Lignum vitae bearing

Fig. 13 – Marine timekeeper H3, by John Harrison, 1740–59

over the next thirty years to discuss progress and agree further payments. As the recipient of ongoing government funding to develop his clocks, Harrison found himself in a unique position. By 1746, he even pleaded that he was so committed to the work that he was 'quite incapable of following any gainfull employment for the support of himself & family'.[24]

Harrison moved to London soon after the Lisbon trial, and within the promised two years he finished his second sea-clock, H2 (Fig. 12). This reflects its London manufacture, since Harrison drew on the capital's skilled clock- and instrument-making workforce for its brass plate, steel springs, engraving and basic finishing. The new clock looked quite different from its predecessor, but was similar in conception, with extensive anti-friction work, a grasshopper escapement and gridiron temperature compensation. It did, however, incorporate some new features, most notably a remontoire, which is a secondary winding mechanism that eliminated variations in the driving force supplied to the escapement and greatly improved the clock's accuracy. Yet H2 never went to trial because Harrison discovered a fundamental flaw that made it susceptible to external motions, and the clock had to be abandoned. A third timekeeper, he assured the Commissioners, would perform better.

Harrison began work on H3 (Fig. 13) in 1740. It was running and being tested within five years, but it was clear from the start that the clock would struggle to keep time to the accuracy desired, forcing him to make changes and adjustments. A drawing completed in the early 1740s, for example, shows a part of the mechanism that Harrison subsequently redesigned (Fig. 14). H3 retained many of the elements of the first timekeepers but also included further innovations. One was a bimetallic strip for temperature compensation, which drew on the principles behind the gridiron pendulum. It consisted of a brass strip and a steel strip riveted together. With one end fixed, the movement of the other adjusted the balance spring as the temperature changed. This idea would come into its own in the twentieth century for switches in kettles, toasters and thermostats. A second innovation was the caged roller-bearing, another anti-friction element, which can be seen as the predecessor of the caged ball-bearing, ubiquitous in complex machinery today.

Despite its novel features, H3 never ran to Harrison's satisfaction, although he laboured on it for nineteen years. As an extraordinary and complex mechanism, however, it attracted attention and praise, gaining Harrison the Royal Society's highest honour, the Copley Medal, in 1749. It features prominently in

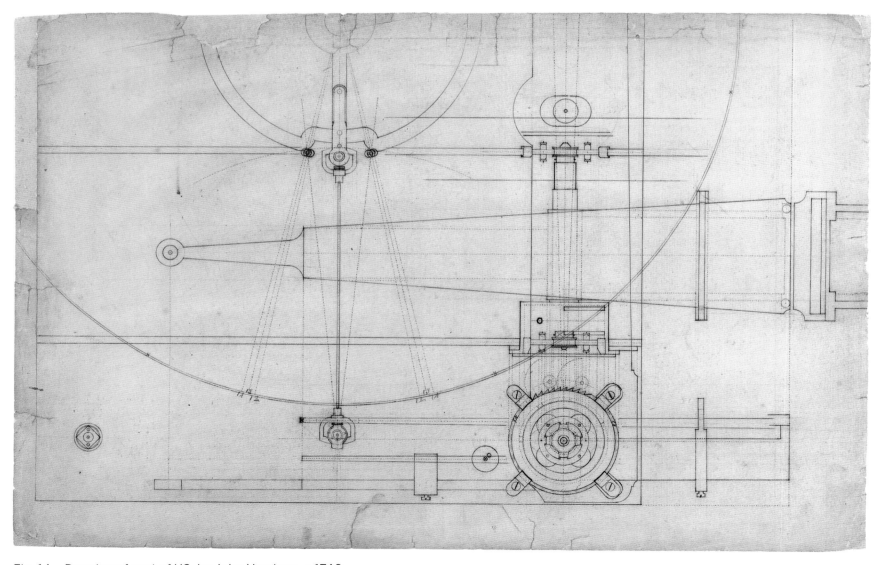

Fig. 14 – Drawing of part of H3, by John Harrison, c.1740

Thomas King's portrait of Harrison (Fig. 15), where it sits in a gimballed case designed to steady it when on a ship (H1 and H2 originally had similar cases). The clock also impressed visitors to Harrison's workshop in Red Lion Square: William Hogarth thought it 'one of the most exquisite movements ever made',[25] while Benjamin Franklin happily paid ten shillings and sixpence for a viewing.

Harrison was already exploring new approaches by the time Franklin visited in 1757, however, and was thinking about watches rather than clocks. After experimenting with various ideas, in about 1751–52 he commissioned John Jefferys to make a watch (which Harrison holds in Fig. 15), with a radically new type of balance. It worked well, so Harrison incorporated it into his fourth longitude timekeeper, H4 (Fig. 16). While it looked like a large pocket watch, H4 was quite different (Fig. 17). The secret can be heard in its rapid ticking, five times a second, since its large balance beats more quickly and with larger oscillations than a typical watch. This contradicted horological orthodoxy,

Fig. 15 – John Harrison, by Thomas King, c.1765–66

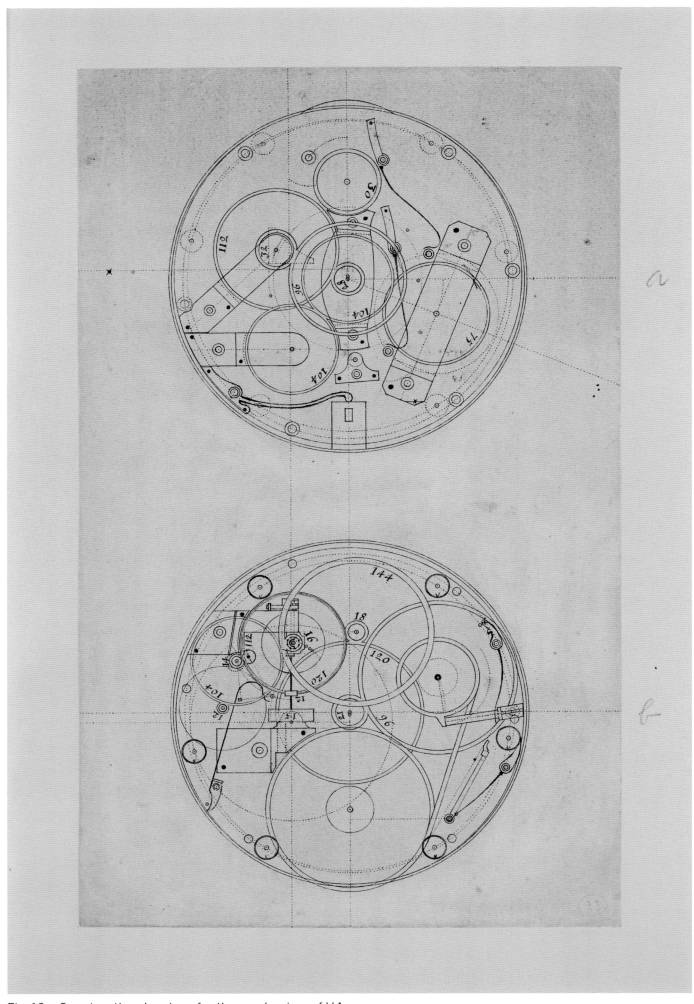

Fig. 16 – Construction drawings for the mechanism of H4, by John Harrison, *c.*1756

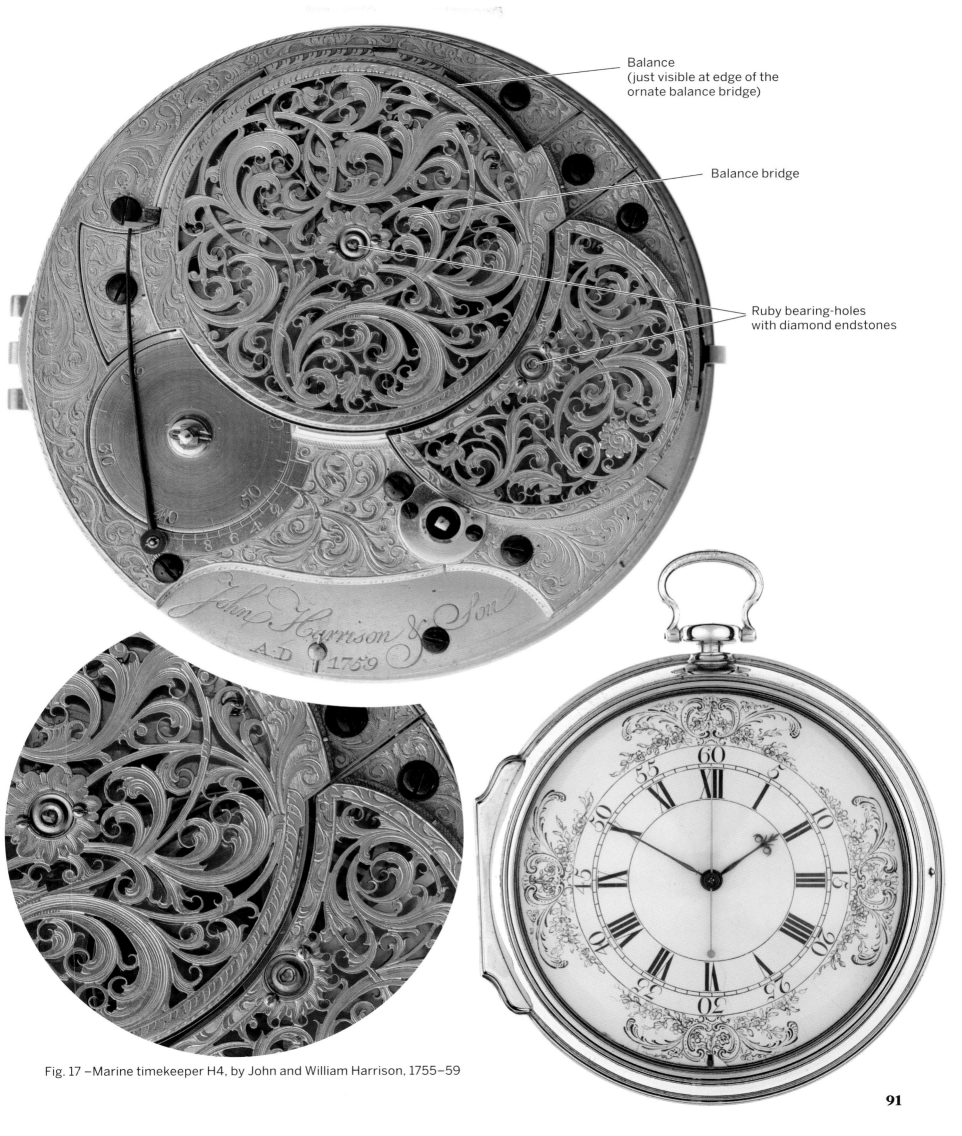

Fig. 17 – Marine timekeeper H4, by John and William Harrison, 1755–59

Fig. 18 – John Hadley, attributed to Bartholomew Dandridge, early 1730s

Fig. 19 – Hadley quadrant or octant, c.1744; although the instrument itself is not signed, there is a handwritten label, now barely legible, signed by George Hadley

which favoured a small, light balance with minimal oscillations. Harrison's thinking was that its rapid pulse would not be affected by a ship's much slower motions and would beat reliably. The only compromise was that H4 needed oil, since Harrison could not miniaturize the anti-friction devices, although he used perfectly shaped jewelled bearings to minimize friction.

The sea-watch was complete by 1759 and, the following July, Harrison suggested that it might accompany H3 on a sea trial. In the end, the continuation of the Seven Years War meant that it was 1761 before the Commissioners gave permission for John's son, William, to prepare for a voyage to Jamaica. Moreover, by the time the *Deptford* sailed with William and H4 on board, John Harrison had decided that H3 would remain in London. The watch's trial seemed to go well. On the way out, William used it to predict an earlier landfall at Madeira than the crew were expecting, so impressing the captain that he asked to buy their next timekeeper. Returning on the *Merlin*, however, William found himself cradling the precious watch in blankets to protect it from tempestuous seas.

It was back in England that trouble began. The Commissioners decided that the test had not, after all, been sufficient. They could not confirm that H4 had succeeded, they said, because the longitude of Port Royal in Jamaica was not properly known. On top of that, they were concerned that its rate, the amount of time it gained or lost each day, had not been agreed beforehand. Despite the Harrisons' protests, the trial was not to be decisive. The Commissioners did concede, however, that H4 was 'an Invention of considerable Utility to the Public' and awarded Harrison £2500, of which £1000 was to be paid after another successful trial.[26] John and William were not mollified.

This was the point when relations between the Harrisons and the Commissioners began to deteriorate. Harrison's friends and supporters began a propaganda campaign of newspaper articles, broadsheets and pamphlets. A petition to Parliament also suggested that he might disclose the watch's principles in return for certain guarantees. A group of 'Commissioners for the Discovery of Mr Harrison's Watch' was duly created but the process broke down and it was decided by all concerned that a second sea trial was the only solution. But other methods had now come to fruition. John Harrison's twenty years as the only serious contender had come to an end. By the 1760s, two rival schemes had emerged to challenge his claim: the measurement of lunar distances (or 'lunars') and observations of the eclipses of Jupiter's satellites. Both would soon be put to test with H4, with all three longitude methods benefiting from the introduction of a new navigational instrument for measuring latitude and local time.

On reflection: the development of the octant

Most longitude methods aimed to establish two local times simultaneously, at the ship and at a reference point of known longitude. Even with a mechanical timekeeper carrying the reference time, local time at the ship had to be found from the Sun or stars, so instruments and techniques for doing this accurately were essential.

The most common method was that of equal altitudes, which involved measuring the altitude of the Sun at some time between three and five hours before midday, and then timing how long it was before the Sun reached the same altitude later in the day. The halfway point between these observations would be local noon. An alternative was to use a single altitude observation of the Sun or a star, although the calculations were more complex. In principle, an instrument for determining latitude could be used for these observations, although greater accuracy than the cross-staff or backstaff could achieve was desirable. For finding longitude by lunar distance, which involved additional observations that were impossible with a backstaff, accuracy was even more critical.

In the 1730s, two men came up with ideas for a more accurate and versatile instrument. In England, it was John Hadley (1682–1744, Fig. 18), who was Vice-President of the Royal Society by the time he addressed a meeting in May 1731. Hadley suggested that optical theory could improve observations by applying the principle of double reflection (that is, using two mirrors in sequence). There was precedent for his ideas; Robert Hooke, Edmond Halley and Isaac Newton had previously devised instruments in which an observer rotated a mirror in order to bring the images of two different objects together – one being reflected in the mirror, the other viewed directly – to measure the angle between them. Edmond Halley had used Newton's instrument on his first Atlantic voyage and was allegedly able to determine longitude 'better then the Seamen by other methods'.[27] Like Newton, John Hadley proposed to use double reflection for indirectly viewing one of the objects; the other being viewed directly. As he demonstrated, this meant that the observations were not greatly affected by the ship's motion – a distinct advantage over other instruments.

As a leading member of the Royal Society who could present his ideas in a solidly theoretical context, Hadley had little difficulty in having his proposal taken seriously and a sea trial in the Thames estuary was organized the following year. It was an august group: Edmond Halley, James Bradley (1693–1762), Professor of Astronomy at Oxford, John Hadley and his brothers Henry and George, who had helped develop the instrument (Fig. 19). Although the observers were 'Persons quite unaccustomed to the Motion of a Ship at Sea' and the Admiralty yacht *Chatham* was small and lively, Hadley reported that good observations were quite possible.[28]

Hadley's claim to priority of invention was nevertheless soon challenged, when the Astronomer Royal received a letter from James Logan, Chief Justice of Pennsylvania in America. Logan described an instrument (Fig. 20) devised by Thomas Godfrey, the son of a maltster, who had trained as a plumber and glazier after being orphaned. Later developing an interest in astronomy and mingling with Philadelphia's shipmasters, Godfrey began working to improve navigational instruments. His double-reflecting instrument was ready in 1730 and tested on the sloop *Trueman*. Having seen it himself, Logan believed it might merit a longitude reward.

Although the two designs differed in the details, Godfrey's plainly used the same principle as Hadley's. After considering the matter, the Royal Society decided that it was a case of independent

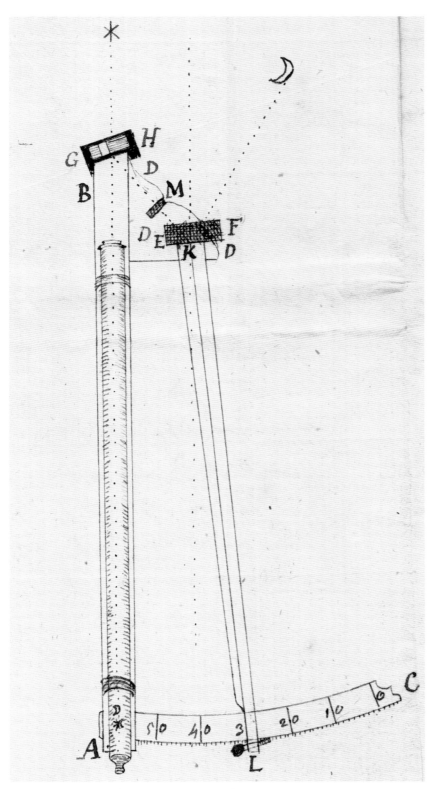

Fig. 20 – Thomas Godfrey's proposal for a double-reflection instrument

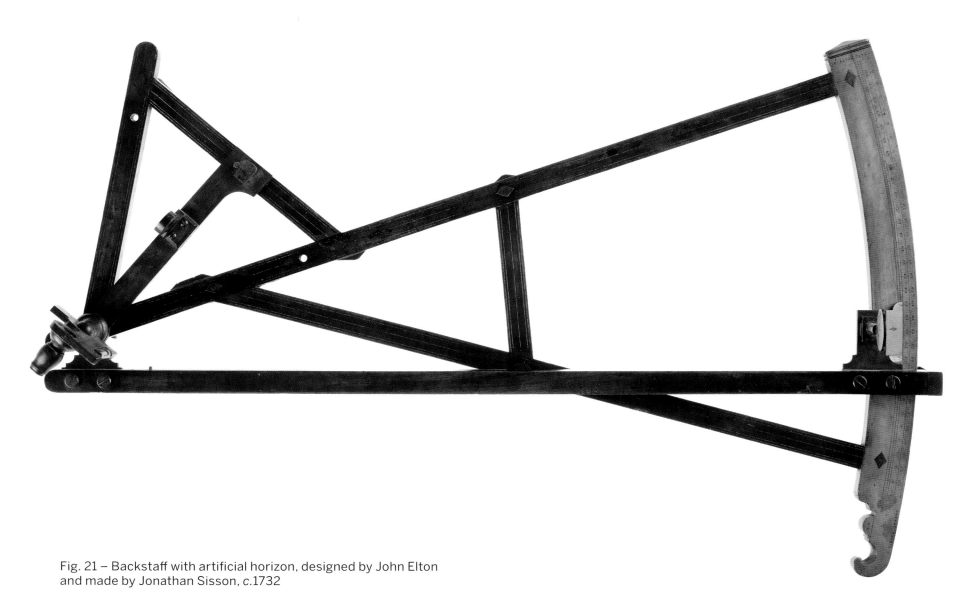

Fig. 21 – Backstaff with artificial horizon, designed by John Elton and made by Jonathan Sisson, c.1732

Fig. 22 – A seaman with an octant, log and line, compass and other equipment, from 'Book of Drafts and Remarks', by Archibald Hamilton, 1763

Fig. 23 – Tobias Mayer, by an unknown artist, mid-eighteenth century

invention. Meanwhile, Hadley took out a patent to bolster his own claim and deter competition in the commercial navigational market. Rivals included new reflecting instruments and improved backstaffs, such as John Elton's design with an artificial horizon for use when the natural one was obscured (Fig. 21).

Instruments to Godfrey's design were manufactured in America for a while, but Hadley's design came to dominate there as it did in England and elsewhere in Europe. Indeed, it would remain popular to the end of the nineteenth century. Later called the octant because the frame was one-eighth of a circle, it was initially known as the Hadley quadrant (since it could actually measure angles up to 90°, or a quarter of a circle). Good connections had ensured that Hadley's name became synonymous with the instrument. And although it found its greatest use for latitude and local time determinations (Fig. 22), the Hadley quadrant would soon be bound up in other attempts to improve astronomical methods for finding longitude.

Mayer, the Moon and Maskelyne

In May 1756, John Elliot, recently promoted to lieutenant in the Royal Navy, was penning a letter home. 'There is no News here worth troubling you with', he wrote, 'only the discovery of the longitude by a Hanovarien … The obs[ervatio]n is simple & easey but the Calculation is extreamly perplexd'.[29] It was a

Fig. 24 – 'Germaniae atque in ea Locorum Principaliorum Mappa Critica', by Tobias Mayer, 1750. Mayer's map compared three sets of co-ordinates – in green, yellow and pink – to show how much they differed

wonderfully understated revelation. The Hanoverian in question was an astronomer named Tobias Mayer (1723–62, Fig. 23). It is said that he had never seen the sea, yet he made possible an astronomical method for determining a ship's longitude. In doing so, he succeeded where Newton had failed, mastering the three-body problem and showing how to predict the Moon's movements.

Mayer's motivation came from a background in cartography, which relied on accurate observations of lunar eclipses and occultations (when one heavenly body obscures another) to measure terrestrial longitudes. Mayer soon realized that existing data was unreliable and drew a map of the German lands to show how uncertain the locations of even the major cities were (Fig. 24). This led him in two directions. First, he wanted to create a better map of the Moon, so that observers could correctly identify specific features to allow more accurate comparisons between data from different places on Earth. So he began a long series of lunar observations, from which he hoped to produce a detailed globe. At the same time, he began to investigate lunar theory and how it might be improved to predict the Moon's motions accurately. This occupied him more and more after his appointment at the University of Göttingen in 1751.

As Newton himself knew, the *Principia*'s theories failed to describe the Moon's motions. Such a conspicuous flaw in the Newtonian system stimulated some of the finest mathematicians of the eighteenth century, including Leonhard Euler, Jean d'Alembert and Alexis Clairaut, to tackle the problem and apply new forms of mathematical analysis in an attempt to model the Sun's effect on the Moon's orbit around the Earth. Such was the crisis that Euler and Clairaut even considered abandoning Newton's inverse square law of gravity to help their models match the observed results. Mayer drew on this work, in particular Euler's techniques, but soon came to believe that part of the problem lay in the astronomical observations on which the theories were based. He confirmed this through a detailed analysis of historical observations and the best modern data from James Bradley, now third Astronomer Royal at Greenwich. This allowed Mayer to apply corrections to the mathematical models and derive an improved theory and new lunar and solar tables, which Bradley found to be impressively accurate.

Mayer did not believe that determining longitude from a ship would ever be possible but, at Euler's urging, he reconsidered and drew up another set of lunar tables and a method for using them at sea. At the same time he designed an instrument for shipboard observations: a circular device that operated on the octant's principle of double reflection but which used repeated measurements around the full circle to reduce the effect of errors in the divided scale (Fig. 25). It all seemed so promising that Euler was soon urging him to submit his ideas to the Commissioners of Longitude, even though Mayer doubted that they would reward a foreigner. Eventually, Mayer gave in and allowed his friends and colleagues to promote his interests in Britain. Thereafter, discussions with the Commissioners passed through diplomatic channels, via Johann David Michaelis, Secretary of Hanoverian affairs in Göttingen, and his cousin William Philip Best, a private secretary of George II. Negotiations were helped by the fact that the King was also the Elector of Hanover.

Discussions of Mayer's ideas were taking place in England by the end of 1754, and James Bradley commissioned the London instrument maker John Bird (1709–76) to make a brass copy of Mayer's repeating circle. This was tested with the lunar tables early in 1757 by Captain John Campbell, who, as a master's mate on the *Centurion* during Anson's traumatic voyage around Cape Horn, knew what issues were at stake. In his tests, Campbell found the circular instrument rather cumbersome and proposed an alternative design comprising just one-sixth of a circle and able to measure up to 120°. This, the first marine sextant, was also made by Bird and was tested by Campbell in 1758–59. It was a success, Bradley reported, concluding that Mayer's tables and a good instrument of this sort (Fig. 26) could determine a ship's longitude to within 1°, a result that brought Mayer's proposal within the limits of the 1714 Act.

While Campbell's trials were promising, they were over short distances during blockade duty as part of the war with France. However, tests over a longer distance were soon underway. One set was carried out by Carsten Niebuhr, astronomer and

Fig. 25 – Mayer's design for a repeating circle, from *Tabulae Motuum Solis et Lunae Novae* (London, 1770)

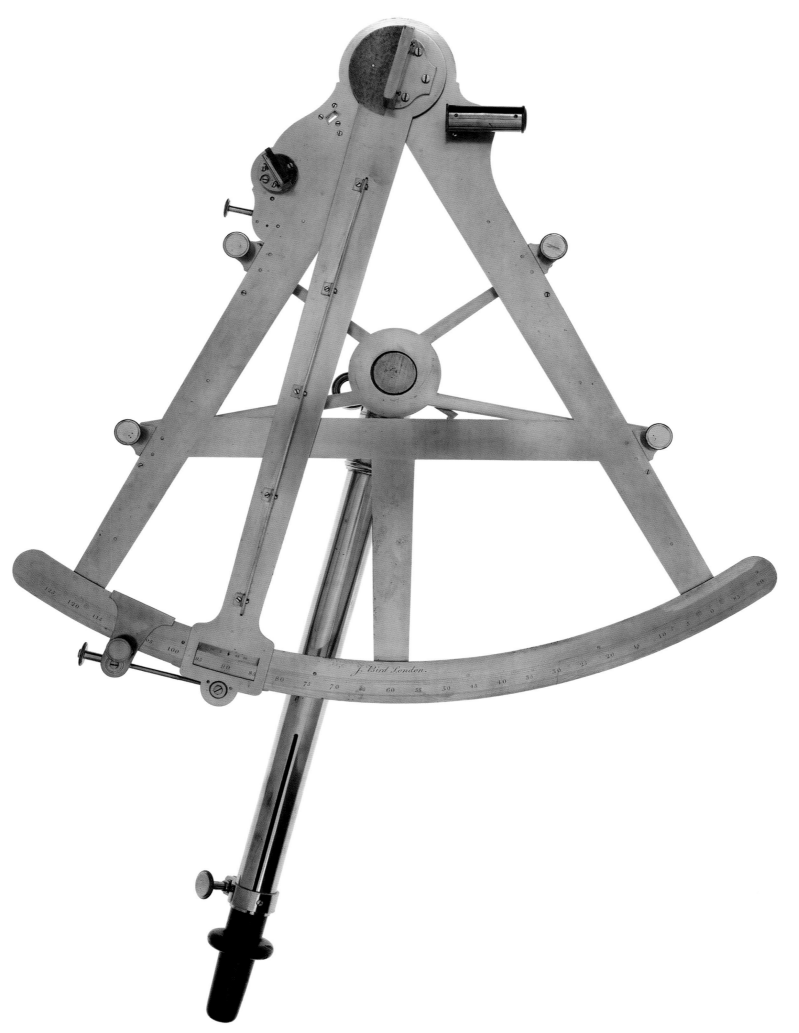

Fig. 26 – Marine sextant, by John Bird, c.1758

Fig. 27 – Nevil Maskelyne, by John Russell, c.1776

cartographer on the Royal Danish Expedition to Arabia in 1761–67. Having studied at Göttingen for the post, Niebuhr had persuaded Mayer to teach him the lunar-distance method and sent back his first results within a few months. Mayer was ecstatic: the observations were 'more accurate than I ever could have hoped for. You have barely taken your first steps at sea and you already can determine longitude better than 80-year-old navigators.'[30]

That same year Mayer's ideas were trialled on a longer British voyage. The opportunity was an expedition to the Atlantic island of St Helena to observe the transit of Venus across the face of the Sun in 1761, a rare event that astronomers hoped to use to determine the distance between the Earth and the Sun and hence the scale of the Solar System. Among other things, the results would help refine the mathematical models of the problematic three-body system. Nevil Maskelyne (1732–1811, Fig. 27), a mathematics graduate from Cambridge who was already working with Bradley, was appointed to make the transit observations and used the voyage to St Helena to test Mayer's ideas. Departing in January 1761, he and his assistant, Robert Waddington, spent eleven weeks on the East India Company ship *Prince Henry* with a Hadley quadrant, Mayer's tables and copies of the French astronomical almanacs, the *Connaissance des Temps*. By the time they anchored, Maskelyne was able to report that his longitude reckoning by lunars was only 1½° in error compared with errors of up to 10° from dead reckoning.

Once he returned from an otherwise frustrating scientific expedition, Maskelyne became a vociferous advocate of lunars. He told the Royal Society that anyone with sufficient time and ability could now determine their longitude at sea. He also published *The*

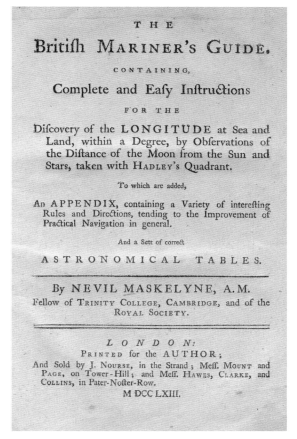

Fig. 28 – *The British Mariner's Guide*, by Nevil Maskelyne (London, 1763)

British Mariner's Guide (Fig. 28), which contained new versions of Mayer's tables and instructions for observing and calculating longitude. He did acknowledge that there might be one objection, however: the 'difficulty and nicety of the calculations', as he gently put it. In fact, the calculations were decidedly lengthy, although he added that they required no knowledge of spherical geometry, only 'care in the computer'.[31] They would be even simpler if pre-computed tables of the Moon's future positions were produced, an idea Maskelyne got from the *Connaissance des Temps* for 1761, which included example tables of this sort and a method for using lunar distances to determine longitude at sea. The method was devised by the French astronomer Nicolas Louis de Lacaille (1713–62), who had used it on a voyage from the Cape of Good Hope in 1754 and published it in full the following year, although his proposal was not fully taken up in France.

Around the same time, Robert Waddington also published a book on lunar distances and began teaching East India Company officers. Lunars had come of age and had active disciples in Britain. Maskelyne would become their most prominent champion and would be a key player in the forthcoming sea trial to the West Indies.

'Erwin's Easy Chair'

While Mayer's negotiations with the Commissioners of Longitude took place through diplomatic channels and international correspondence, other more public strategies could be successful. Christopher Irwin was a master of these. Indeed, most of what is known of him, including the fact that he was from Roscommon in Ireland, comes from his self-publicity in newspapers and other publications.

Irwin came to the attention of London readers in the late 1750s, when he began promoting a scheme for finding longitude from Jupiter's satellites. His proposal was for a marine chair with counterweights hanging underneath, to hold the observer steady as they viewed the satellites through a telescope. The idea of building such a contraption predated even the discovery of Jupiter's satellites (Fig. 29). In that sense, Irwin was only repeating what had gone before, but he did incorporate the latest telescopic innovations. More notable was his use of newspapers to convince the public and the Commissioners that his design worked.

In 1759, Irwin persuaded the Navy to lay on shipboard trials of the chair. Within days, *The London Magazine* reported that he had discovered the longitude and printed a certificate from the 'brave lord Howe', a hero from the Seven Years War, testifying that the chair could be used to determine longitude to within fifteen miles. The report added that the king's brother was much taken with it and that his mathematical teacher had cried, 'This will do, this will do'.[32] Irwin was presumably responsible for other flattering reports in London's press, including a letter from a 'Francis Drake', who applauded 'the excellency of the design, and the masterliness of the execution' and urged the government to reward its inventor.[33] Not everyone was taken in: the *Busy Body* ran a fictional autobiography of a would-be projector, whose scheme had obtained 'a pompous certificate, signed by an admiral, three captains of men of war, and a mathematical Professor, who repeated to me these flattering words, *this will do, this will do*.'[34]

Whoever the authors, the publicity caused quite a stir. In October, a Swedish astronomer named Bengt Ferrner saw the chair at the workshop of Jeremiah Sisson, who had it mounted above his house with a hole through the roof for the counterweight. Ferrner was not convinced it would work: while it was certainly 'artful and comfortable', a ship's movements would probably disturb it too much.[35] Around the same time, one of Tobias Mayer's supporters scribbled a concerned note to Göttingen about the rival invention, which was being called 'Erwin's Easy Chair'.[36]

Irwin's publicity drive seems to have impressed the Commissioners, since they met him during 1762 and, swayed by Lord Howe's report, offered £500 for further experiments. Irwin made sure the papers reported it, one praising 'the intense labour and unrebukeable perseverance of a fellow subject indued with great candor and modesty'.[37] The next year, Irwin told the Commissioners that he had made further improvements but wanted more money. Eventually, they agreed another £100 for him to join the sea trials being discussed with Harrison. All was set for the Easy Chair to go to the West Indies.

Trial by water

With three longitude methods ready by 1763, the end of the Seven Years War allowed a full trial to go ahead and the Commissioners finalized the details in August. The destination was to be Barbados, with Maskelyne appointed as the astronomer in charge, assisted by Charles Green, an astronomical assistant at the Royal Observatory. On the way out, Maskelyne and Green were to test Irwin's marine chair and Mayer's newest tables, which his widow had sent after his death in 1762. Once at Barbados, they were to determine the island's longitude by observations of Jupiter's satellites in order to assess the two astronomical methods

Fig. 29 – A sixteenth-century design for a marine observing chair, from *Le Cosmolabe*, by Jacques Besson (Paris, 1567)

Fig. 30 – 'View of Bridgetown and part of Carlisle Bay in the Island of Barbadoes', by Edward Brenton, late eighteenth century

and the performance of H4, which would travel separately with William Harrison.

Maskelyne, Green and Irwin departed England on the *Princess Louisa* in September 1763, arriving in Bridgetown (Fig. 30) in early November. The two methods had fared very differently. The astronomers had made many successful lunar-distance observations, Maskelyne noting that the results of his final set were within ½° of the truth. 'My friend Irwin's machine', he confided to his brother, however, 'proves a mere bauble'.[38] Even in calm waters, Jupiter and its satellites moved too rapidly across the field of view, although Green did manage at least one successful observation. For the second half of the voyage, they abandoned the chair and focused on the Moon.

William Harrison left the following March on the *Tartar*. The watch performed well throughout and impressed the ship's captain. William had reason to feel confident, therefore, until he came ashore in mid-May. According to his later account,

> he was told that Mr. Maskelyne was a Candidate for the Premium for discovering the Longitude and therefore they thought it was very odd, that he should be sent to make the Observations to Judge another Scheme Mr. Maskelyne having declared in a very Public manner that he had found the Longitude himself ...[39]

Harrison claimed that Maskelyne was so discombobulated when challenged on the matter that his observations became sloppy and worthless. Maskelyne's actual thoughts are unknown, but it is worth noting that Harrison junior was hardly an objective advocate for his father. Nor is there evidence that the young astronomer sought a reward, since it was Mayer's work that was on trial, not his. Nonetheless, the incident was symptomatic of deteriorating relations between the Harrisons, the Commissioners and the astronomical community represented by Maskelyne.

Harrison and Maskelyne returned to England separately the same year. The Commissioners sent the results for processing and were ready to consider them on 9 February 1765, Maskelyne having been appointed as fifth Astronomer Royal only the day before. No doubt to the consternation of the Harrisons, he was now a Commissioner too.

There was much to discuss at the meeting. In the light of Maskelyne's testimony that lunar distances could find longitude to within 1°, the Commissioners recommended that Mayer's widow receive up to £5000 in recognition of her husband's work. They also confirmed that John Harrison's watch had kept time within the most stringent limits of the 1714 Act, its error being just 39.2 seconds or 9.8 miles (15.8 km) at the latitude of Barbados. They pointed out, however, that he

> hath not yet made a discovery of the Principles upon which the said Timekeeper is constructed, nor of the method of carrying those principles into Execution, by means whereof other such Timekeepers might be framed of sufficient correctness to find the Longitude at sea ... whereby the said Invention might be adjudged practicable and usefull in terms of the said Act & agreeable to the true Intent & meaning thereof.[40]

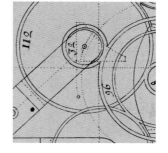
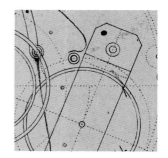

Their recommendation was that Parliament award Harrison £10,000 when he demonstrated the principles of the watch, with the remaining £10,000 (less payments already made) to be awarded once it was proved that 'his method will be of common & general Utility'; in other words, once it was shown that other makers could produce similar timekeepers.[41] This would set the terms of a series of increasingly heated debates with the Harrisons, who considered that the full reward was already due under the terms of the 1714 Act and that the Commissioners had unfairly changed the rules.

The recommendations went before Parliament and became law in a new Longitude Act of 10 May 1765. This confirmed the conditional payments to Harrison, but awarded only £3000 to Mayer's heirs, as well as £300 to Leonhard Euler for theoretical work that underpinned the development of accurate lunar tables. In addition, the Act proposed that £5000 might be paid to anyone who improved Mayer's tables in the future and instructed the Commissioners to begin publishing a nautical almanac. The trials were over. It was up to the Commissioners to bring the new methods into practice.

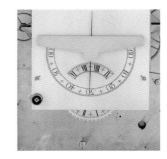

CHAPTER 4
MAKING LONGITUDE WORK

We have employed proper persons to compute a Nautical Almanac and Astronomical Ephemeris ... which will greatly contribute to make the Lunar Tables constructed by the late Professor Mayer ... more generally useful ...

Foreword to *The Nautical Almanac and Astronomical Ephemeris, for the Year 1767* [1]

The 1765 Longitude Act had two main functions. First, it provided a statutory process through which to reward key individuals, such as John Harrison, who had developed successful ways to measure longitude. Second, it provided a way of investing in and supporting, for the foreseeable future, the two methods of finding longitude that had proved to be the most successful: lunar-distance and timekeeper. This strategy was not a way of hedging bets, or of supporting one method until the other was available for wider use, but was based on the view that they were complementary. Astronomical methods, though tricky to use, were the only means of finding longitude if it were lost and of checking that a timekeeper was working sufficiently well. They were also considered the more accurate methods, under ideal observing conditions and with many repetitions.

The new Act also recognized something of which the Commissioners had become ever more aware: the terms of the 1714 Act were insufficient for selecting a method that could simply be rolled out across the Navy. The Commissioners were conscious that they were dispensing large amounts of public money and were at risk of having little to show for it. The need for further work and investment had to be explained carefully to Members of Parliament, who might baulk at continued spending and fail to grasp why a successful trial did not mean the problem was solved.

For the lunar-distance method, successful implementation meant continually publishing predictive astronomical tables to ease the burden of calculation for the user. The 1765 Act therefore stated that to make Mayer's 'Lunar Tables more generally useful', the Commissioners should

> cause such Nautical Almanacks, or other useful Tables, to be constructed, and to print, publish, and vend ... any Nautical Almanack or Almanacks, or other useful Table or Tables, which they ... shall, from time to time, judge necessary and useful, in order to facilitate the Method of discovering the Longitude at Sea ... [2]

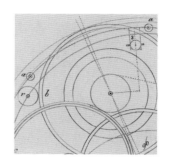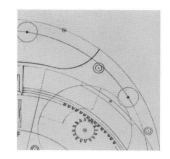
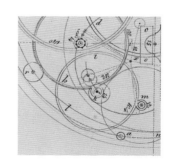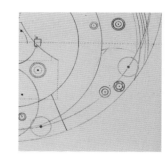

The bare construction of one single machine can never be said to discover the longitude.

From an anonymous letter to *The Gentleman's Magazine* (1765)[3]

Fig. 1 – The Royal Observatory from the south-east, unknown artist, c.1770

106

Nevil Maskelyne, now Astronomer Royal, was in a position to focus work at the Royal Observatory on the production of observations that would serve this purpose. On top of this, he directed the considerable calculating work required to turn raw observational data into usable tables, published as the *Nautical Almanac*. This was, as suggested in the *British Mariner's Guide*, to be modelled on the tables for 1761 that had been computed by the French astronomer Nicolas Louis de Lacaille for the *Connaissance des Temps*.

The *Almanac* also supported the use of timekeepers at sea. Anticipating a future in which they were more plentiful, it provided information on how to take and process observations to establish local time, compare this to a timekeeper and check its going (that is, its speed and regularity), at sea or on land. However, there was much more to be done to make this method widely available. One watch – one very expensive, complex and slowly constructed watch – was of little use: the Commissioners had to know that Harrison's work could be replicated. Harrison was, after all, now in his seventies and it was a very real possibility that his secrets might die with him.

By this time, reflecting their more regular transaction of business, the Commissioners had become widely known as the Board of Longitude. Their collective decision, at a meeting on 9 February 1765, was that, although the watch had passed the trial, they could not be confident that it had solved the problem. Harrison would only be paid 'upon his producing his Timekeeper to certain persons to be named by this Board & discovering to them, upon oath, the principles & manner of making the same'.[4] Harrison vehemently disagreed, believing that, since his watch had performed as specified in the 1714 Act, he should immediately receive the full reward.

Ultimately, new watches would have to be made, with cheaper and simpler constructions. Maskelyne and the Royal Observatory were central to this process. The Astronomer Royal assessed

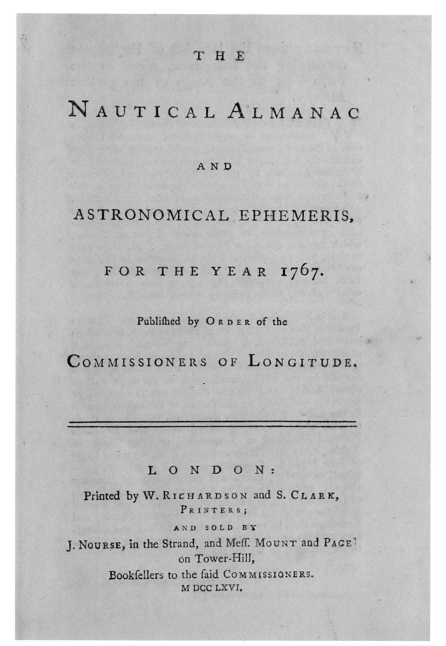

Fig. 2 – The title page and one of the tables from *The Nautical Almanac and Astronomical Ephemeris* for 1767

ideas, dealt with makers, encouraged the spread of horological knowledge, oversaw trials of timekeepers at the Observatory and encouraged their use at sea. He and the Board also facilitated the exchange of knowledge about longitude methods across national borders. In continental Europe, astronomers and clockmakers had been engaged in similar researches, with their own networks of patronage and financial support. And in France, in particular, effective marine timekeepers were beginning to be produced.

Maskelyne and the *Nautical Almanac*

Working on behalf of the Board of Longitude, Maskelyne quickly created a system for producing the *Nautical Almanac and Astronomical Ephemeris* (Fig. 2). Within months of his arrival at Greenwich in March 1765, he had established the layout and content of each *Almanac*, providing information useful to astronomy, navigation and cartography, initially for a year ahead, and later for several years in advance. Work got under way, and the data for 1767 were ready to be printed by the end of 1766. As the Astronomer Royal told his brother, then living in India,

> There will be 12 pages in every month. All the lunar calculation for finding the longitude at sea by that method will be ready performed: & other useful & new tables added to facilitate the whole calculation; so that the sailers will have little more to do than to observe carefully the moon's distance from the sun or a proper star, which are also set down in the ephemeris, in order to find their longitude.[5]

It was not quite as simple as this sounded but the 'ready performed' calculation considerably shortened the process of establishing a reference time, in this case Greenwich time. The tables would also enable the establishment of longitudes on land, by observations of Jupiter's satellites, and facilitate calculation of latitude and local time.

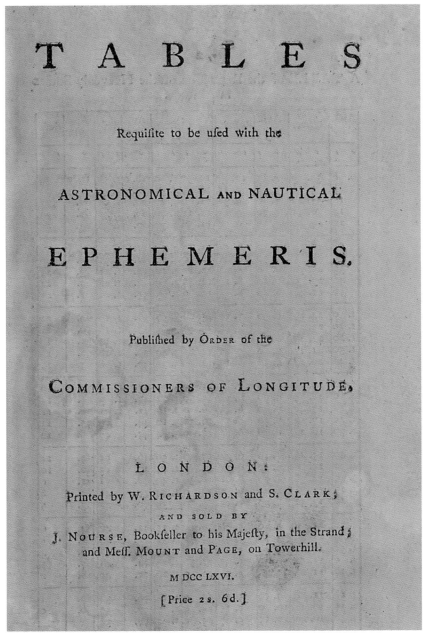

Fig. 3 – The title page and a page of tables from the *Tables Requisite to be used with the Astronomical and Nautical Almanac* (London, 1766)

Maskelyne noted also that, to complete this endeavour, 'The board of longitude have engaged persons to compute' this 'very complete nautical & astronomical ephemeris'.[6] These were Israel Lyons, George Witchell, William Wales and John Mapson, and they were appointed on 13 June 1765. To ensure the accuracy of the calculations, these 'computers', as they were known, worked independently in pairs, so that the work of each member of a pair could be checked against that of the other. Lyons and Witchell worked on the data for January to June 1767, while Wales and Mapson covered July to December. The pattern of work for the next several decades was set when Richard Dunthorne was appointed 'comparer' to check the accuracy of each pair's work and correct as necessary. Each computer and comparer was paid £70 for a year's calculations, though this was soon raised to £75, probably because the slow, painstaking nature of the work became apparent.

Dunthorne also worked with Maskelyne to produce another volume, the *Tables Requisite to be used with the Astronomical and Nautical Almanac* (Fig. 3), which provided additional information that did not require annual updating. It aimed to be clear and

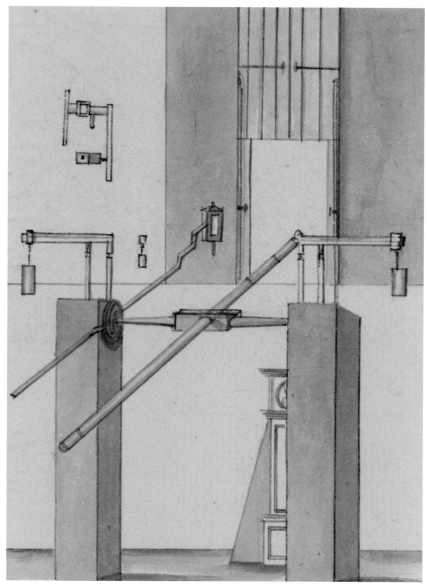

Fig. 4 – The John Bird transit instrument and astronomical regulator made by George Graham in the Royal Observatory, Greenwich, drawing by John Charnock, about 1785 (detail)

Fig. 5 – Nevil Maskelyne's 'observing suit' of padded silk, probably designed to keep him warm during cold, night-time observations, about 1765

Fig. 6 – Nevil Maskelyne, by Louis François Gérard van der Puyl, 1785

helpful, explaining how to take observations, what instruments to use and how to apply the tables. Both this and the first edition of the *Almanac* were available from the printer on 6 January 1767. While the Board seems to have overestimated the market – of the 1000 *Almanacs* and 10,000 *Tables Requisite* printed, 242 and 6992 respectively remained unsold by 1784 – this was the start of a very long series. Maskelyne oversaw and edited forty-nine issues of the *Almanac* and three editions of the *Tables Requisite*. The publication of the *Almanac* continues to this day.

Maskelyne's dedication and management were essential to the success of the process, although he did not receive additional income for adding the role of *Almanac* editor to that of Astronomer Royal. Other, paid, individuals were required to sustain the project. Closest to home were the successive astronomical assistants at Greenwich. Their priority was to assist with observing the Moon, measuring its height and timing the moment that it crossed the meridian of the fixed observatory instruments, such as the John Bird transit instrument and astronomical regulator made by George Graham (Fig. 4).

The transits of the Sun and planets were also observed when possible, as were lunar and solar eclipses and other conjunctions of the Sun, Moon, planets, satellites and stars. All of these could be used to establish and refine the measurement of the difference in longitude between Greenwich and any other location at which the same observations had been made. Stellar observations were largely limited to thirty-one (later thirty-six) stars that lie near the celestial equator, the imaginary circle in the sky that lies in the same plane as the Earth's Equator. The stars chosen were bright enough to be viewed with a telescope in daylight and were spread across the sky. These stars were used to correct the observatory clocks and became known as 'clock stars'. An overlapping group of ten bright stars, in addition to the positions of the Sun and Moon, were observed for tabulation in the *Nautical Almanac*.

Observations were generally carried out by astronomer and assistant together: Maskelyne's padded silk 'observing suit' shows signs of significant wear (Fig. 5). As Astronomer Royal, he worked under rules that he had helped the Royal Society establish in the year before his appointment. The rules stipulated that he should not be absent for more than a few days without permission and that he must ensure that either he or his assistant were resident at all times. It had also become a requirement for copies of the observations carried out at the Observatory to be submitted annually to its Board of Visitors. They were later printed, thus beginning the regular series of *Greenwich Observations*. Maskelyne's portrait shows him with the first edition of 1776, containing ten years' observations (Fig. 6).

It was largely Maskelyne's assistants who processed, or reduced, the observations into data ready for publication. It was tedious and poorly paid work and seems to have led to a high turnover of staff, so Maskelyne asked that their pay be increased. The rise from £26 to £86 must have been welcome, yet, as Maskelyne himself wrote, the role still required an individual with good eyes for the telescope, good ears for the ticking of the clock, a good constitution, and the ability to work on calculations several hours a day, get up at night and 'bear confinement'.[7] The role was often seen as an opportunity for excellent observational and mathematical training rather than as a long-term position.

Of the twenty-four individuals who were assistants during Maskelyne's forty-six year tenure, many went on to become observers elsewhere, or teachers of mathematics, astronomy and navigation. Often benefiting from Maskelyne's patronage, they helped to spread knowledge of the new longitude methods. Several former assistants either regularly or irregularly took on computational work for the *Nautical Almanac* and the Board of Longitude. Such work also added to the income of others with mathematical ability, good literacy, focus and a neat hand, including teachers, astronomers, ministers, surveyors and instrument makers. Such people had useful skills but often precarious careers.

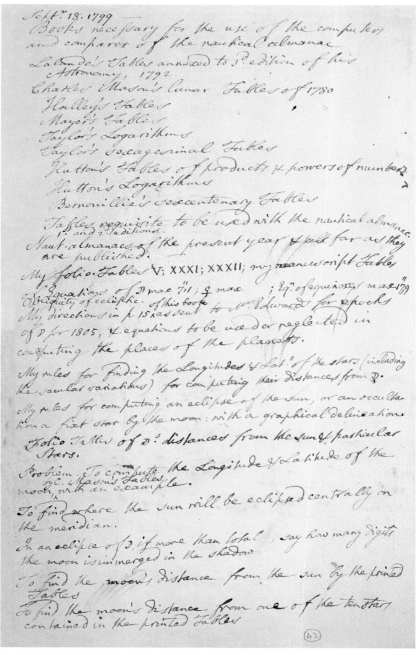

Fig. 7 – List of reference books required by the computers, compiled by Nevil Maskelyne in 1799

Most of the methods they used had been outlined in the *British Mariner's Guide*, although the computers and comparer undoubtedly helped refine them for production on a larger scale. Instructions were sent out by post, as were the work, queries, corrections and completed calculations. The computers were also sent a large number of astronomical and mathematical tables as reference tools (Fig. 7) that they needed to consult in order to process a set of observations. These books of tables belonged to the Board of Longitude, and Maskelyne was always careful to ensure their safe return if one of its employees ceased work.

While most of the computers were known or personally recommended to Maskelyne, they were geographically scattered. They were able to work from home and, often, in between other professional activities. Payment was organized by month of the *Almanac* computed, and Maskelyne became skilled at anticipating the rate of work and monetary requirements of his different computers, keeping a careful track of work as it was parcelled out (Fig. 8).

In December 1767 it was decided that the *Almanac* should be published three years in advance in order to provide for longer voyages. Maskelyne was authorized to recruit more computers, two of whom were Joseph Keech and Reuben Robbins. They evidently found that they had taken on more work than they anticipated, for they took the shortcut of copying each other's calculations rather than working the same data independently. The comparer, Malachy Hitchins, proved his ability to detect errors or, in this case, suspicious regularity. Their cheating was revealed and the pair were dismissed and told to recompense Hitchins for his wasted time. Thereafter, Maskelyne made a virtue of the geographical spread of his computers, by making sure that each computer in a pair lived at a distance from each other. He obviously forgave Keech and Robbins, as they did further computing, but they never again worked on the same month's data.

Maskelyne employed a total of thirty-five computers and comparers. By 1789, though, he was using only four – Mary Edwards, Henry Andrews, Joshua Moore and Malachy Hitchins – as the process became streamlined and the computers adept. The system and these loyal employees ensured that accuracy was largely maintained. Problems crept in after the deaths of Hitchins, who had been comparer for forty years, and Maskelyne, especially under the less strictly controlled regime of his successor, John Pond (1767–1836).

Many of the stories of individual computers make fascinating reading but Mary Edwards stands out as one of the few female contributors to this narrative. Even her name might have been lost, had she not outlived her husband. John Edwards was

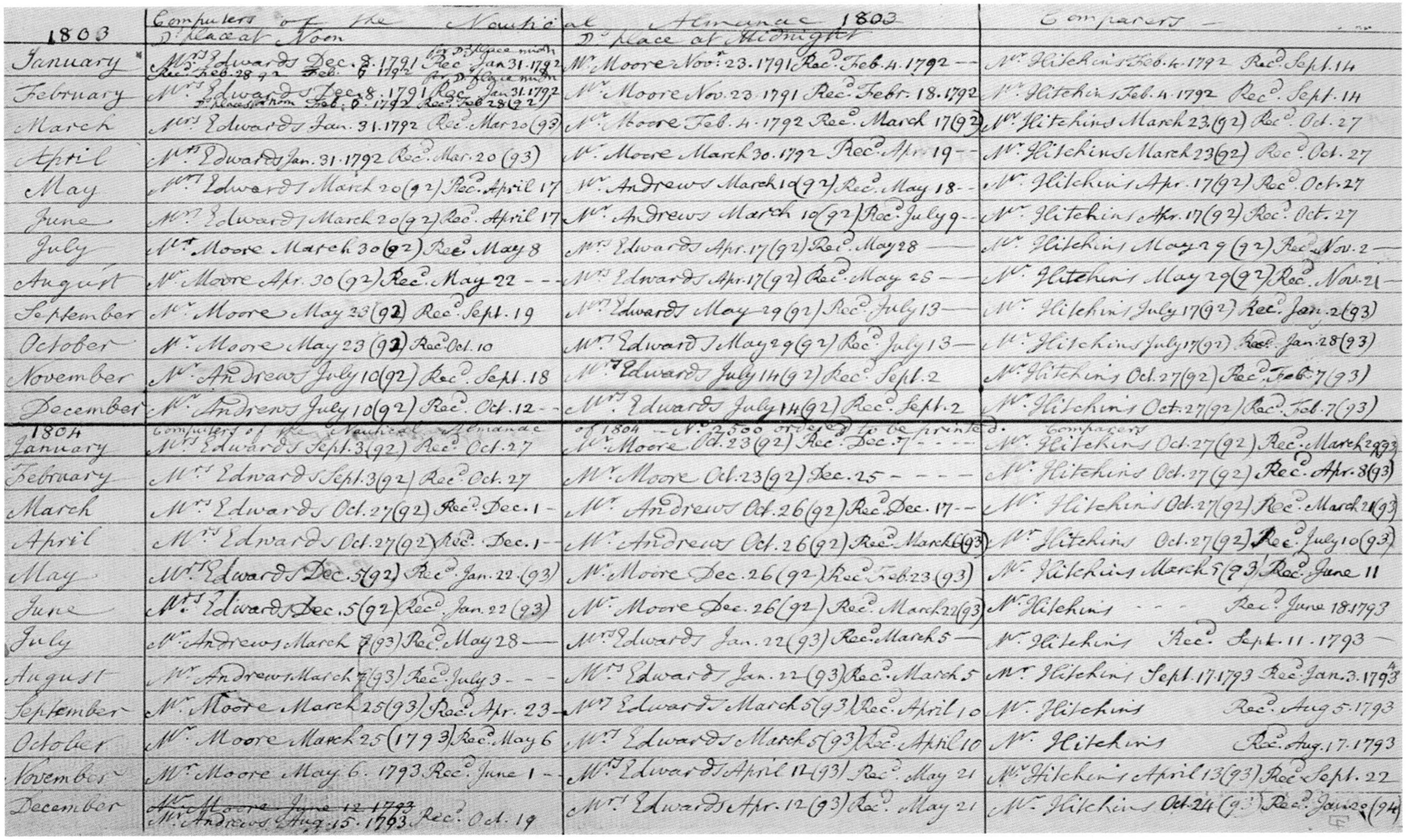

1803	Computers of the Nautical Almanac 1803 — D. place at Noon	D. place at Midnight	Comparers
January	Mrs Edwards Dec. 8. 1791 Rec. Jan 31. 1792	Mr Moore Nov. 23. 1791 Rec. Feb. 4. 1792	Mr Hitchins Feb. 4. 1792 Rec. Sept. 14
February	Mrs Edwards Dec. 8. 1791 Rec. Jan 31. 1792	Mr Moore Nov. 23. 1791 Rec. Febr. 18. 1792	Mr Hitchins Feb. 4. 1792 Rec. Sept. 14
March	Mrs Edwards Jan. 31. 1792 Rec. Mar 20 (93)	Mr Moore Feb. 4. 1792 Rec. March 17 (92)	Mr Hitchins March 23 (92) Rec. Oct. 27
April	Mrs Edwards Jan. 31. 1792 Rec. Mar. 20 (93)	Mr Moore March 30. 1792 Rec. Apr. 19	Mr Hitchins March 23 (92) Rec. Oct. 27
May	Mrs Edwards March 20 (92) Rec. April 17	Mr Andrews March 10 (92) Rec. May 18	Mr Hitchins Apr. 17 (92) Rec. Oct. 27
June	Mrs Edwards March 20 (92) Rec. April 17	Mr Andrews March 10 (92) Rec. July 9	Mr Hitchins Apr. 17 (92) Rec. Oct. 27
July	Mr Moore March 30 (92) Rec. May 8	Mrs Edwards Apr. 17 (92) Rec. May 28	Mr Hitchins May 29 (92) Rec. Nov. 2
August	Mr Moore Apr. 30 (92) Rec. May 22	Mrs Edwards Apr. 17 (92) Rec. May 25	Mr Hitchins May 29 (92) Rec. Nov. 21
September	Mr Moore May 23 (92) Rec. Sept. 19	Mrs Edwards May 29 (92) Rec. July 13	Mr Hitchins July 17 (92) Rec. Jan. 2 (93)
October	Mr Moore May 23 (92) Rec. Oct. 10	Mrs Edwards May 29 (92) Rec. July 13	Mr Hitchins July 17 (92) Rec. Jan. 28 (93)
November	Mr Andrews July 10 (92) Rec. Sept. 18	Mrs Edwards July 14 (92) Rec. Sept. 2	Mr Hitchins Oct. 27 (92) Rec. Feb. 7 (93)
December	Mr Andrews July 10 (92) Rec. Oct. 12	Mrs Edwards July 14 (92) Rec. Sept. 2	Mr Hitchins Oct. 27 (92) Rec. Feb. 7 (93)
1804	Computers of the Nautical Almanac	of 1804 — N. 2500 ordered to be printed.	Comparers
January	Mrs Edwards Sept. 3 (92) Rec. Oct. 27	Mr Moore Oct. 23 (92) Rec. Dec. 7	Mr Hitchins Oct. 27 (92) Rec. March 29 (93)
February	Mrs Edwards Sept. 3 (92) Rec. Oct. 27	Mr Moore Oct. 23 (92) Rec. Dec. 25	Mr Hitchins Oct. 27 (92) Rec. Apr. 8 (93)
March	Mrs Edwards Oct. 27 (92) Rec. Dec. 1	Mr Andrews Oct. 26 (92) Rec. Dec. 17	Mr Hitchins Oct. 27 (92) Rec. March 21 (93)
April	Mrs Edwards Oct. 27 (92) Rec. Dec. 1	Mr Andrews Oct. 26 (92) Rec. March 6 (93)	Mr Hitchins Oct. 27 (92) Rec. July 10 (93)
May	Mrs Edwards Dec. 5 (92) Rec. Jan. 22 (93)	Mr Moore Dec. 26 (92) Rec. Feb. 23 (93)	Mr Hitchins March 5 (93) Rec. June 11
June	Mrs Edwards Dec. 5 (92) Rec. Jan. 22 (93)	Mr Moore Dec. 26 (92) Rec. March 22 (93)	Mr Hitchins Rec. June 18 1793
July	Mr Andrews March 7 (93) Rec. May 28	Mrs Edwards Jan. 22 (93) Rec. March 5	Mr Hitchins Rec. Sept. 11. 1793
August	Mr Andrews March 7 (93) Rec. July 3	Mrs Edwards Jan. 22 (93) Rec. March 5	Mr Hitchins Sept. 17. 1793 Rec. Jan. 3. 1794
September	Mr Moore March 25 (93) Rec. Apr. 23	Mrs Edwards March 5 (93) Rec. April 10	Mr Hitchins Rec. Aug 5. 1793
October	Mr Moore March 25 (1793) Rec. May 6	Mrs Edwards March 5 (93) Rec. April 10	Mr Hitchins Rec. Aug. 17. 1793
November	Mr Moore May 6. 1793 Rec. June 1	Mrs Edwards April 12 (93) Rec. May 21	Mr Hitchins April 13 (93) Rec. Sept. 22
December	~~Mr Moore June 12 1793~~ Mr Andrews Aug. 15. 1793 Rec. Oct. 19	Mrs Edwards Apr. 12 (93) Rec. May 21	Mr Hitchins Oct. 24 (93) Rec. Jan. 20 (94)

Fig. 8 – List by Nevil Maskelyne of work allocated to computers and comparers in 1791–93 for the 1803 *Nautical Almanac*

Fig. 9 – Date and signature on the upper plate of John Harrison's H4 sea-watch

a Shropshire clergyman who lived near Maskelyne's sister Margaret, Lady Clive, and probably became known to them after making his own telescope mirrors. He used his astronomical interests and mathematical skills to add significantly to his family income, receiving payment for work on six months' worth of each *Almanac* from 1773 to 1784. However, it seems likely that he handed on much or most of the calculating work to his wife, for when he died the accounts moved seamlessly from 'John Edwards' to 'Mary Edwards'.

It was unusual for women at this time to have had the mathematical knowledge and training to work as computers, but Mary may have been taught by her husband and must have had a natural aptitude. She, in turn, gained enough experience to help teach new computers, including her daughters, and after the death of Hitchins she was also able to take on work as a comparer. She worked fast and had a very low rate of errors. Earning under her own name, she increased her workload, computing twelve months of each year's *Almanac*. As half of the whole computing power required annually, this was more than any of the other computers.

After Pond took over as editor of the *Nautical Almanac* in 1811, Edwards suddenly found her computing work being reduced and her comparing work stopped altogether. She petitioned the Board, which acknowledged her good work but did not reinstate her to the more prestigious position of comparer. She was, however, effectively paid a premium for twelve months' calculations while only doing eight. Edwards died in 1815 but her daughter Eliza continued as a computer until 1832, when the work of the *Almanac* was consolidated and ceased to be the cottage industry in which the Edwards had specialized. As in other fields in the nineteenth century, such women had their earning power removed by changes to the organization of workplaces.

The 'Discovery of Mr Harrison's Watch'

The *Nautical Almanac* was intended to help make the new longitude methods 'practicable and useful at sea'. This phrase in the 1714 Act provided the Commissioners of Longitude with much needed wriggle-room: it was decided that only with additional information could Harrison's watch fulfil the intent of the Act. To deal responsibly with public money, the full £20,000 reward would only be paid when the Board was satisfied 'that his method will be of common & general Utility'.[8]

As a petition put to Parliament in Harrison's name back in 1762 had suggested, a 'discovery' of the watch should be made 'in such manner as that other workmen may be able to Execute the same so that it may in a short time become serviceable to this Kingdom and to all who use the Sea'.[9] Yet Harrison still hoped that he could be paid the full £20,000 and retain a monopoly on future manufacture. He told the Board that they could either give him £800 to make two more watches that would let him claim the second £10,000, or they could advance it and allow him to set up a factory and employ and train other workmen. The Board refused both options.

The disclosure of the mechanism of Harrison's fourth longitude timekeeper, H4 (Fig. 9 and Chapter 3, Fig. 17), was to be

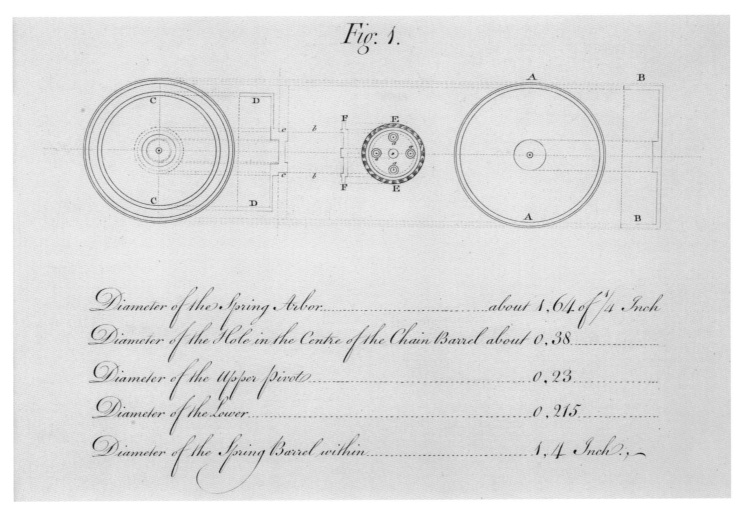

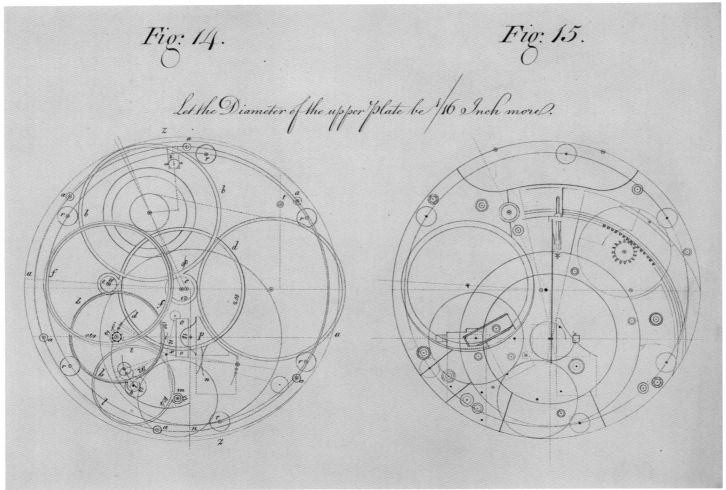

Fig. 10 – Drawings from *The Principles of Mr. Harrison's Timekeeper* (London, 1767)

Fig. 11 – Marine timekeeper K1, by Larcum Kendall, 1769

witnessed by a reappointed Commission for the Discovery of Mr Harrison's Watch, overseen by Maskelyne. These were experts whom the Admiralty could trust and individuals who might understand and reproduce the mechanism: mathematicians John Mitchell and William Ludlam; instrument maker John Bird; and watchmakers Thomas Mudge, William Matthews and Larcum Kendall. The 'full discovery' was to include explanation of the tools used, details of the methods of tempering the metals, and spoken, written and drawn descriptions of the mechanism. The Commissioners were aware of their responsibility to follow the description well enough to be able to report and, perhaps, make such a watch themselves.

The disclosure took place in August 1765 at Harrison's house and workshop in Red Lion Square. After six days of explanation and demonstration, the Commissioners declared themselves satisfied. Interestingly, though, four of the six – Ludlam, Mitchell, Kendall and Bird – would later express doubts over the adequacy of the explanation or the practicality of the watch. Ludlam suggested in print that the methods were not readily transferable, that the disclosure was incomplete, that Harrison had made adjustments by trial and error rather than by following a method, and that the watch's temperature adjustment was inadequate.

Harrison handed his timekeepers over to the Admiralty, much against his will but as a long-standing condition of the rewards he had received so far. They would be taken to Greenwich for further

Fig. 12 – Marine timekeepers, by John Arnold, c.1771

trials – the beginning of the Observatory's enduring role as a place for trialling, storing and checking marine timekeepers. In 1766, H4 was subjected to a ten-month trial against the Observatory's astronomical regulator. The watch was wound daily, usually by Maskelyne, sometimes by his assistant, but always in the presence of an officer of Greenwich Hospital, where the key for one of the two locks on the box was kept.

It did not perform well over this period as a whole, with its rate (the amount of time it gained or lost each day) appearing erratic. The trial appeared to show that, statistically, H4 was likely, but not certain, to perform successfully over the course of a six-week voyage. It also suggested that the corrections made to compensate for its rate would have to be frequently adjusted over the course of a longer voyage, especially in cold weather, making the watch less generally practical. Maskelyne found that it 'cannot be depended upon' to keep the longitude within a degree on a voyage to the West Indies, nor to keep it within half a degree for more than a fortnight, and it could be much worse if conditions were unfavourable. Nevertheless, he concluded that it would be 'a useful and valuable invention' that, used 'in conjunction' with lunars to check whether and by how much its rate changed over the course of the voyage, 'may be of considerable advantage to navigation'.[10]

Harrison attacked Maskelyne's integrity in the conduct of the trial, his care of the timekeepers and his claims for lunar distances, accusing him of being 'deeply interested' in the lunar scheme. The clockmaker emphasized that, if the rate of the watch was regularly recalculated and compensated for, any six-week period of the trial, save those in extreme cold or where the watch had not been kept horizontal, would yield a result that was within the limits of the 1714 Act. Maskelyne did not disagree but this did not answer the greater question of whether the watch had solved the problem more generally: could it be used on longer voyages and in extreme conditions? Could another be made?

Harrison was, in the meantime, making a second watch but continued to complain about his treatment and about having to work under the new Act. His viewpoint was irreconcilable with that of the Commissioners. Ultimately he bypassed the Board and turned to Parliament and George III. The Board of Longitude also looked elsewhere. This did not mean they were prejudiced against or disbelieving of timekeepers as a viable means of finding longitude. Their investment in Harrison had been huge and they were not interested in letting it prove a dead end.

Making more timekeepers

As Maskelyne informed his brother in 1766, shortly after the arrival of H4 at the Observatory,

> The board of longitude are also desirous to encourage the making of watches after Mr. Harrison's method. They have engaged a person to make one. I have had the drawings engraved here under my eye, & shall publish them in a short space of time.[11]

The person engaged was Larcum Kendall (1719–1790), whom William Ludlam believed was 'more likely to make such a Watch than any body of the Trade that he knows of'.[12] Kendall had been apprenticed to John Jefferys and may have helped make some parts of H4, as well as having been present at its disclosure. He proposed to make a replica in two years for £450. It would be as identical as possible, 'But I will not be answerable for its keeping time'.[13] This proviso was inserted on Ludlam's private suggestion that he make no guarantees that it would perform as Harrison's had: 'it might be good fortune in that case, & the thing never happen again'.[14]

The Commissioners did not, however, wish to rely only on Harrison and Kendall. Early on, Kendall had been instructed to

Fig. 13 – *Horloge marine* no. 8, by Ferdinand Berthoud, 1767

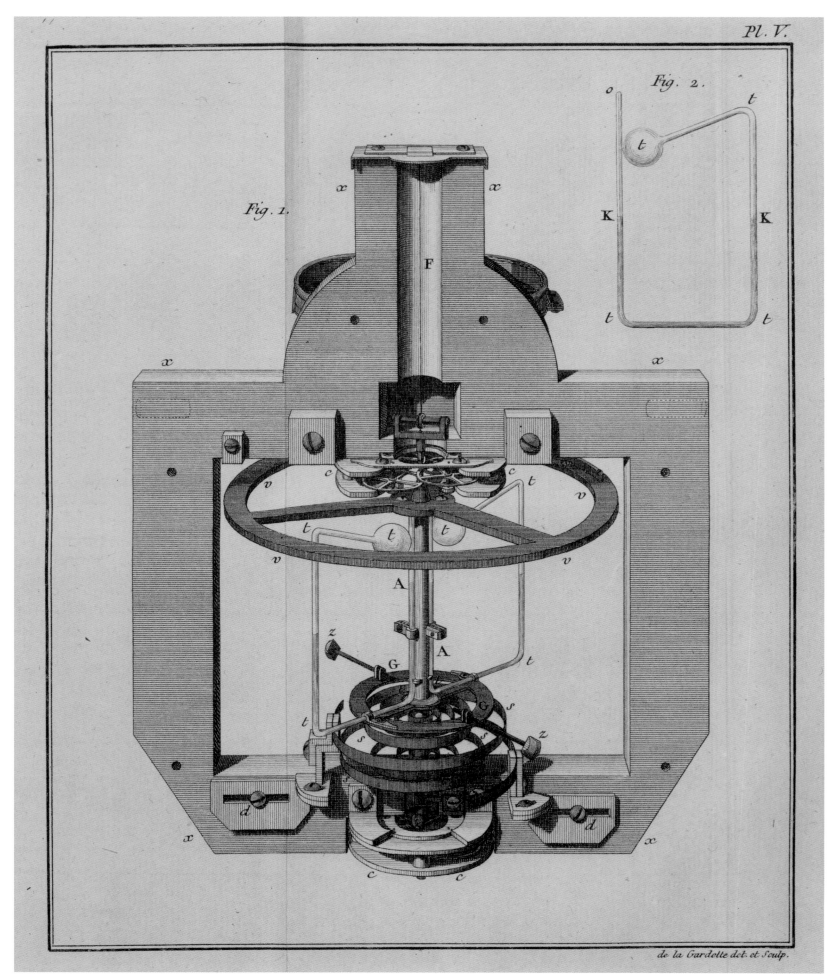

Fig. 14 – The balance assembly of Pierre Le Roy's *montre marine* (marine watch) in Le Tellier, *Journal du Voyage de M. le Marquis de Courtanvaux* (Paris, 1768). The assembly incorporated temperature compensation with two adjustable balance springs below

let not only the Harrisons see H4, if required, but also any other 'ingenious persons' who expressed an interest.[15] Another essential means of disseminating Harrison's ideas was the publication of his drawings and descriptions in the 1767 *Principles of Mr. Harrison's Timekeeper* (Fig. 10), edited by Maskelyne. Five hundred copies were printed, with translations into French and Danish appearing with astonishing speed. It was hoped that this would spark imitation and innovation in the design of timekeepers. The Board, therefore, continued to offer rewards for improved timekeepers, as well as other methods.

Kendall completed his copy of Harrison's watch, today known as K1 (Fig. 11), and presented it to the Board on 13 January 1770. It was pronounced a faithful copy and Kendall was given both the £450 and a further reward of £50. He was asked to take on the task of instructing other workmen to make the necessary parts for further copies. On 26 May 1770, however, he demurred. There were 'so many difficultys' in the undertaking, he said, 'I am well assured in my own mind; I shall not be able to do the publick any service'.[16]

What the Board wanted, Kendall understood, was a means by which 'a Watch might in a short time be made for a reasonable price'. His view was that 'it would be many years (if ever) before a watch of the same kind, with that of Mr Harrisons; could be afforded for Two hundred Pounds'. He explained that making more replicas would:

> never answer [the Board's ends] so effectually; as to make the watch become of publick utility. for unless a reduction; of the expensive parts of the Watch can be effected; & some method contrived, to facilitate the adjusting; the watch would still come to so high a price; as to put it far out of the reach of purchase for general use.[17]

Kendall told the Board that, instead of producing another copy, changes to Harrison's design might allow him to make a watch that was just as good but less than half the price. This watch, K2 (see Chapter 5, Fig. 14), was delivered two years later. He explained his simplifications, adjustments and improvements, and claimed 'my Watch is reduced to much greater simplicity than Mr Harrison's and may be much easier adjusted; therefore may be made at a less expence'.[18] Kendall expressed confidence in its performance but also indicated that he had come up with an even simpler design. The Board agreed to let him try, and K3 was delivered for £100 in 1774 (see Chapter 5, Fig. 17).

By the 1770s, the Commissioners were thoroughly focused on Kendall's simplifications and the work of another maker, John Arnold (1736–1799). Although Arnold first came before the Board in 1770, it appears that Maskelyne had already singled him out as a talented workman who might be put to good use, for he had sent him a copy of *The Principles of Mr. Harrison's Timekeeper*. At the same meeting at which Kendall said he could make a watch for £200, Arnold had shown his own timekeeper, claiming that with further work it might be made for just sixty guineas (£63). The Board was impressed enough to advance him £500 to continue with his work.

Arnold was clearly influenced by the published description of H4 in *Principles*, and certain details were copied directly from its engravings. He had also taken up hints from elsewhere, including Maskelyne's report that Harrison thought his timekeeper would be better placed in a wooden box than a silver case. Arnold's early marine timekeepers therefore look more humble than H4 or K1, although they included precision work (Fig. 12). They also contained a new type of detached escapement, known as a detent escapement – an idea probably borrowed from France – which was effective in reducing interference between wheels and balance, which Harrison himself had said was desirable. From 1771, after tests at the Royal Observatory, Maskelyne instigated plans to have these watches tested on James Cook's second Pacific voyage, alongside K1.

The desire of the Commissioners to publish and share information contrasted sharply with Harrison's instinct to protect his methods. When the explanation of his watch had first been mooted in 1763, members of the Board had even invited the French to send nominated experts to join the witnesses. The mathematician Charles-Étienne Camus and the clockmaker Ferdinand Berthoud (1727–1807) were selected and arrived in London. Since this planned disclosure did not take place, they attempted to get information directly from Harrison, making contact through the French astronomer Jérôme Lalande, who was then resident in London.

Fig. 15 – Marine timekeeper, by William Snellen, *c*.1775, and probably influenced by Harrison's work

Harrison refused, both in 1763 and again when Berthoud returned to London in 1766. Industrial espionage, including attempts to transfer the secrets of the London instrument trade overseas, was a feature of the period, and artisans jealously guarded their secrets and access to their workshops. However, Harrison's clocks were not a typical case of intellectual property. Berthoud was, therefore, able to get a description of the watch directly from Thomas Mudge when they met at the house of Count Bruhl, who also acted as interpreter. Harrison complained but Mudge explained that he considered this openness to be desirable: 'I thought it my duty to do it, and that it was the Intention of the Board I should do so'.[19] He had also shared the information with ten to twelve English workmen.

Berthoud's personal qualities stood in marked contrast to Harrison's and were, it would seem, distinctly more amenable to and in accord with the mores of the Paris and London scientific academies. Berthoud, Camus and Lalande were all elected as fellows of the Royal Society in 1764, and it is notable that Berthoud's election certificate, as well as mentioning that he had been chosen to witness the much negotiated 'discovery' of H4, stresses the fact that he had published several books describing his work on marine timekeepers, had willingly shown his clock to the Académie des Sciences and had deposited it with their Secretariat.

Berthoud did not gain sufficient information to make watches like Harrison's, despite his conversation with Mudge and possession of *Principles*. It is possible that the tight-knit clockmaking circles of London shared an understanding of Harrison's work and skills that the clockmakers of Paris could not. Certainly Harrison's descriptions were no help. The clockmaker Charles-Pierre d'Evreux de Fleurieu reported that 'he had veiled his works so as to let them be seen without that it is possible to copy them'.[20] It is likely, though, that Berthoud would not have followed Harrison's path even if he could. The quality and accuracy of his marine timekeepers has sometimes been questioned but his focus, during a long and successful career, was on producing a good number of sufficient timekeepers rather than a small number of exquisite ones (Fig. 13).

Berthoud was one of two clockmakers working on marine timekeepers with support from the French king, government and Académie. The other, his senior, was Pierre Le Roy (1717–1785), who, over time, succeeded in producing timekeepers incorporating three elements that horologists consider essential to the later development of the chronometer: a detached escapement, temperature-compensated balance and isochronous balance spring (that is, one that transmitted its driving force at absolutely regular intervals) (Fig. 14). However, his clock designs managed to be both too large and insufficiently robust for practical use, contrasting with Berthoud's focus on utility. While the success of Harrison's watch had depended on a high-energy balance that produced long, fast oscillations, Le Roy's clocks had a large balance, with short, slow oscillations, and even his *petites rondes* – small timekeepers within gimballed boxes, produced from about 1771 – had, like most watches, a low-energy balance that was too easily disturbed by motion.

Neither Le Roy's nor Berthoud's timepieces owed much to Harrison's designs. Just as the French had shown the way with the publication of astronomical tables for navigation, they had a long record of research and investment in the timekeeper method. Le Roy and Berthoud had begun their research in the 1750s, were trialling timekeepers in the 1760s and received rewards in the 1770s, expending significant energies along the way on battling each other over who had invented what and who was the more deserving of patronage, titles and commissions. Le Roy seems to have been the more difficult character, although he is credited for the originality of his work. Berthoud managed his negotiations better and was always ready to share and publish his ideas. Equally, he borrowed from others, but his focus on strength, usability, simplicity of manufacture and ease of repair were arguably more important. He was able to produce seventy timekeepers, many of which were used on voyages of exploration.

Knowledge of British and French work in marine timekeepers circulated widely in European maritime spheres.

Fig. 16 – Marine timekeeper H5, by John Harrison, completed 1770

Harrison's *Principles* may, for example, have influenced a watch made by the Dordrecht instrument maker William Snellen in the 1770s (Fig. 15). This watch contains elements reminiscent of Harrison's work, although its construction is very different. In many countries, however, attempts to manufacture marine timekeepers did not take off until the late eighteenth or nineteenth century and even then sometimes unsuccessfully. The British makers had made a head start and, for some time yet, those wishing to purchase instruments of the highest quality would largely look to London.

The end of the Harrison affair

Although he had spoken of working on timekeepers in the plural, John Harrison, now in his late seventies, succeeded in making only one more marine watch, known as H5 (Fig. 16). Taking a new initiative, the Harrisons made contact with Stephen Demainbray (1710–1782), astronomer at George III's private observatory at Kew, and requested a trial there. Demainbray raised the issue with the King in January 1772, and William Harrison was invited to Windsor. Long after the event, John Harrison's grandson, also called John Harrison, reported that George III had then exclaimed, 'these people have been cruelly treated' and, 'By God, Harrison, I will see you righted!'[21]

The meeting led to a trial of H5 at the Kew Observatory from May to July 1772. As with the sea and Greenwich trials of H4, the watch was placed in a box with more than one lock, to prevent tampering. In this case it was Demainbray, William Harrison and the King who were to be present for the comparisons with the regulator clock at noon each day. After a false start, which has rather unconvincingly been attributed to placing it too close to some lodestones, the watch performed impressively. The recorded daily rate of variation over the whole ten weeks, which saw fairly constant temperatures, has been averaged out at less than a third of a second per day (Fig. 17).

With one last crack at the Board of Longitude, Harrison sent a communication that was discussed at their meeting of 28 November 1772. This covered much old ground but added the new results for H5, showing '(as he alledges) it went very considerably within the … nearest Limits prescribed' by the 1714 Act.[22] He hoped thereby to receive the remainder of the large reward. The Board called in William Harrison and stated that they saw no reason to change the agreed approach, which required an official trial. At this same meeting they took up Kendall's offer to construct his third watch and heard that John Arnold was keen to have two more of his timekeepers tested. K1 and three Arnold timekeepers were concurrently undergoing

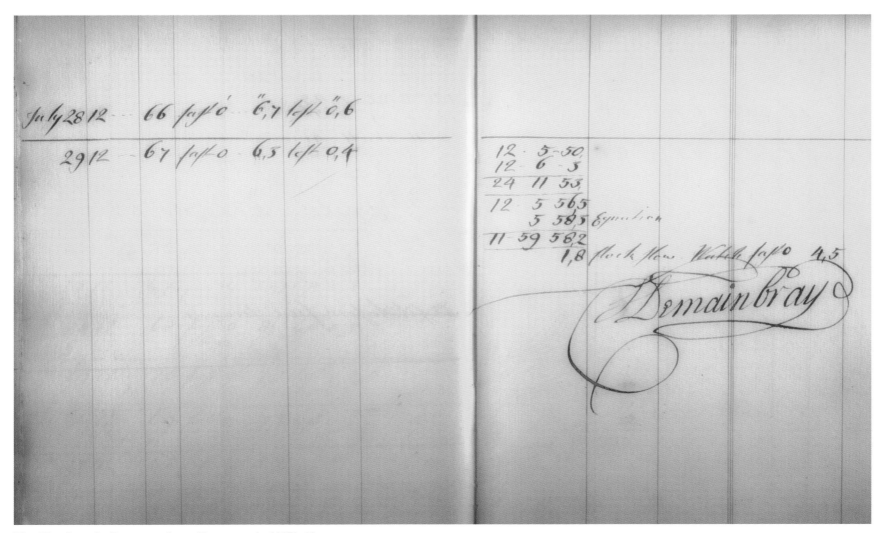

Fig. 17 – Concluding page from the record of H5's Kew Observatory trial, by Stephen Demainbray, 1772 (detail)

the most rigorous of tests on board Captain Cook's ships *Resolution* and *Adventure*.

William Harrison, however, continued to have the ear of the King, as suggested by a 1773 letter in which he claimed 'to lay before His Majesty every Tuesday everything which I have done, and I do not write one word or take one step without acquainting him with it'.[23] It was thus presumably through the King's influence that they approached the Prime Minister, Lord North and the Secretary to the Treasury, John Robinson. The Harrisons' petitions asked for 'bare justice'. Their long-standing paranoia about Maskelyne remained, and they raised objections to the role that he would play in further tests as he had 'every Tie of Interest' and 'stands pledged to crush the Invention'.[24]

A petition was presented to Parliament on 2 April 1773, again blaming the Commissioners and asking for the second £10,000. Meanwhile, William Harrison was asked by the Board of Longitude why he and his father would not make another timekeeper, or consider submitting H5 to trial under 'two or more persons (to whom you have no objection)'. The Board minutes note his reply as: 'Loss of time, Expense attending it, Uncertainty of reward afterwards and I think I can employ my time better'.[25] The breakdown in relations was complete.

Harrison's petition was debated on 27 April. When Lord North outlined the Board's concerns, the Whig leader, Edmund Burke, replied with a rousing defence of this 'most ingenious and able mechanic', who had, 'according to the verdict of the whole mechanical world, done more than ever was expected'.[26] Harrison was, however, advised that the Board had acted legally and so he substituted a new petition that simply called for generosity and assistance. Lord North reported that the King supported this petition and, on 1 July 1773, an Act awarded Harrison a sum not exceeding £8750 for his decades of dedication, 'as a further Reward & Encouragement over & above the Sum already received by him'.[27] This decision took the total amount of money that Harrison had received from government above £20,000 but it also ensured that this payment was less than the £10,000 potentially offered under the 1765 Act.

Doubt therefore remains, and opinions are divided, over whether it can be said that Harrison had finally won the large reward offered under the 1714 Act. Harrison could stress that, if you added all large and small rewards, he had received at least £20,000 and so the large reward had been paid. However, on other occasions, he also argued that much of this money had been for expenses and that he should still be given the additional £1250 allowed under the 1765 Act. It seems clear that Parliament did not want to be seen to be overruling the decisions of the Board or the legally enshrined conditions of the 1765 Act and, to be sure, £8750 was a very large sum for an expression of the nation's gratitude for Harrison's work.

Many have argued that Harrison should have received £20,000 fair and square after the Barbados trial and before the passing of the game-changing 1765 Act. The Board had chosen to worry about whether or not that trial had shown that Harrison possessed a practicable and useful method that would benefit the public. The question remains: were the Commissioners acting unfairly, being over-conscientious or doing their public duty? Was the 'Harrison method' that was on trial simply a *single timekeeper* that proved capable of doing the job, or was it *the means of making a successful marine timekeeper*? If the latter, success could only be proved by making more of them.

The Board of Longitude had, as much as possible, extracted information from Harrison and stepped away from his ongoing complaints. However justified Harrison may have been in his interpretation of the 1714 Act, the Commissioners were concerned about the accountability of their actions and were at the end of their collective tether in dealing with him. Different expectations about the need for openness played their part, although it appears that there were problems communicating with Harrison that did not arise with other clockmakers. Harrison had difficulty expressing himself and his son is widely regarded as having been an unpleasant individual. Their increasing suspicion, together with the Commissioners' sense that they could now get better results by looking elsewhere, led to the end of a long relationship. Nevertheless, Harrison died a wealthy man on 24 March 1776 (Fig. 18).

Fig. 18 – Medallion portrait of John Harrison by James Tassie, c.1776

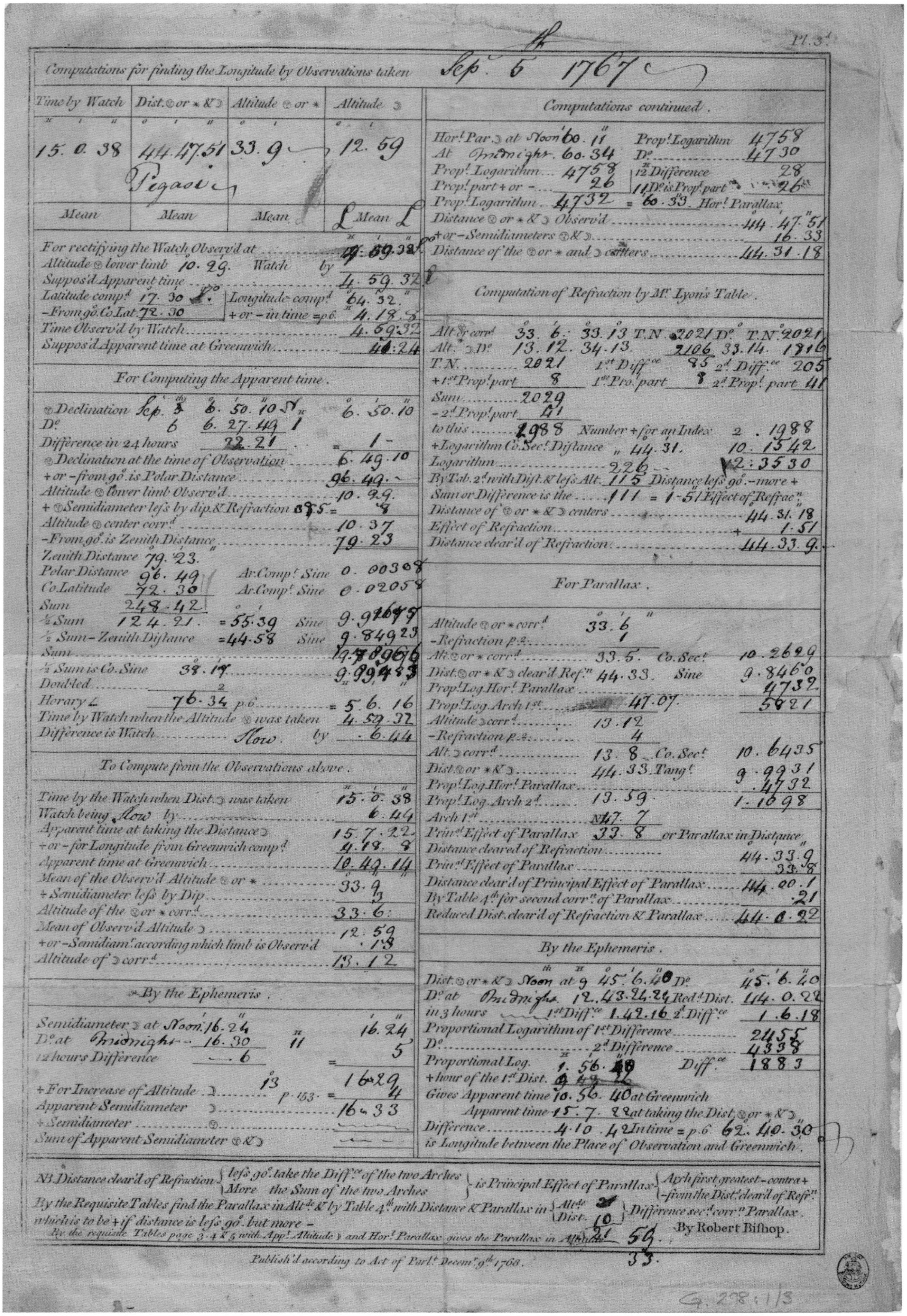
Pl. 3.

Computations for finding the Longitude by Observations taken Sep. 6 1767

Time by Watch	Dist. ☉ or ✶ & ☽	Altitude ☉ or ✶	Altitude ☽
15. 0. 38	44. 47. 51	33. 9	12. 59
	Pegasi		
Mean	Mean	Mean	Mean

For rectifying the Watch Observ'd at ... 4. 59. 32
Altitude ☉ lower limb 10. 29. Watch by
Suppos'd Apparent time ... 4. 59. 32
Latitude comp.d 17. 30 N. Longitude comp.d 64. 32.
-From 90. Co. Lat. 72. 30 + or - in time = p. 6. 4. 18. 8
Time Observ'd by Watch ... 4. 59. 32
Suppos'd Apparent time at Greenwich ... 9. 18. 24

For Computing the Apparent time.

☉ Declination Sep. 5 6. 50. 10 N. 6. 50. 10
D.o 6 6. 27. 49
Difference in 24 hours 22. 21 = 1 -
☉ Declination at the time of Observation ... 6. 49. 10
+ or - from 90. is Polar Distance ... 96. 49.
Altitude ☉ lower limb Observ'd ... 10. 29.
+ ☉ Semidiameter less by dip & Refraction ... 8
Altitude ☉ center corr.d ... 10. 37
-From 90. is Zenith Distance ... 79. 23
Zenith Distance 79. 23.
Polar Distance 96. 49 Ar. Comp.t Sine 0. 00308
Co. Latitude 72. 30 Ar. Comp.t Sine 0. 02058
Sum 248. 42
½ Sum 124. 21. = 55. 39 Sine 9. 91675
½ Sum - Zenith Distance = 44. 58 Sine 9. 84923
Sum ... 19. 78966
½ Sum is Co. Sine 38. 17 9. 89483
Doubled ... 2
Horary ∠ 76. 34 p. 6 = 5. 6. 16
Time by Watch when the Altitude ☉ was taken 4. 59. 32
Difference is Watch ... Slow by 6. 44

To Compute from the Observations above.

Time by the Watch when Dist. ☽ was taken 15. 0. 38
Watch being Slow by ... 6. 44
Apparent time at taking the Distance ☽ 15. 7. 22
+ or - for Longitude from Greenwich comp.d 4. 18. 8
Apparent time at Greenwich ... 10. 49. 14
Mean of the Observ'd Altitude ☉ or ✶ ... 33. 9
+ Semidiameter less by Dip ... 3
Altitude of the ☉ or ✶ corr.d ... 33. 6.
Mean of Observ'd Altitude ☽ ... 12. 59
+ or - Semidiam.r according which limb is Observ'd 13
Altitude of ☽ corr.d ... 13. 12

By the Ephemeris.

Semidiameter ☽ at Noon 16. 24 16. 24
D.o at Midnight 16. 30 11
12 hours Difference 6 = 5
+ For Increase of Altitude ☽ ... 13 p. 153 = 16. 29 4
Apparent Semidiameter ☽ ... 16. 33
+ Semidiameter ☉
Sum of Apparent Semidiameter ☉ & ☽

Computations continued.

Hor.l Par. ☽ at Noon 60. 11 Prop.l Logarithm 4758
At Midnight 60. 34 D.o 4730
Prop.l Logarithm ... 4758 12 Difference 28
Prop.l part + or - ... 26 11 D.o is Prop.l part 26
Prop.l Logarithm ... 4732 = 60. 33. Hor.l Parallax
Distance ☉ or ✶ & ☽ Observ'd ... 44. 47. 51
+ or - Semidiameters ☉ & ☽ ... 16. 33
Distance of the ☉ or ✶ and ☽ centers ... 44. 31. 18

Computation of Refraction by M.r Lyon's Table.

Alt. ✶ corr.d 33. 6. 33. 13 T.N. 2021 D.o T.N. 2021
Alt. ☽ D.o 13. 12. 34. 13. 2106 33. 14. 1816
T.N. ... 2021 1.st Diff.ce 85 2.d Diff.ce 205
+ 1.st Prop.l part 8 1.st Pro. part 8 2.d Prop.l part 41
Sum ... 2029
- 2.d Prop.l part 41
to this ... 1988 Number + for an Index 2. 1988
+ Logarithm Co. Sec.t Distance 44. 31. 10. 1542
Logarithm ... 226 2. 3530
By Tab. 2.d with Dist. & less Alt. 115 Distance less 90. - more +
Sum or Difference is the ... 111 = 1. 51 Effect of Refrac.n
Distance of ☉ or ✶ & ☽ centers ... 44. 31. 18
Effect of Refraction ... + 1. 51
Distance clear'd of Refraction ... 44. 33. 9.

For Parallax.

Altitude ☉ or ✶ corr.d 33. 6
- Refraction p. 4. 1
Alt. ☉ or ✶ corr.d ... 33. 5. Co. Sec.t 10. 2629
Dist. ☉ or ✶ & ☽ clear'd Ref.n 44. 33. Sine 9. 8460
Prop.l Log. Hor.l Parallax ... 4732
Prop.l Log. Arch 1.st ... 47. 07. 5821
Altitude ☽ corr.d ... 13. 12
- Refraction p. 4. ... 4
Alt. ☽ corr.d ... 13. 8. Co. Sec.t 10. 6435
Dist. ☉ or ✶ & ☽ ... 44. 33. Tang.t 9. 9931
Prop.l Log. Hor.l Parallax ... 4732
Prop.l Log. Arch 2.d ... 13. 59. 1. 1098
Arch 1.st ... 47. 7
Prin.l Effect of Parallax 33. 8 or Parallax in Distance
Distance clear'd of Refraction ... 44. 33. 9
Prin.l Effect of Parallax ... 33. 8
Distance clear'd of Principal Effect of Parallax ... 44. 00. 1
By Table 4.th for second corr.n of Parallax ... 21
Reduced Dist. clear'd of Refraction & Parallax ... 44. 0. 22

By the Ephemeris.

Dist. ☉ or ✶ & ☽ Noon at 9 45. 6. 40 D.o 45. 6. 40
D.o at Midnight 12 43. 24. 24 Red.d Dist. 44. 0. 22
in 3 hours 1.st Diff.ce 1. 42. 16 2.d Diff.ce 1. 6. 18
Proportional Logarithm of 1.st Difference ... 2455
D.o 2.d Difference ... 4338
Proportional Log. 1. 56. 40 Diff.ce 1883
+ hour of the 1.st Dist. 9
Gives Apparent time 10. 56. 40 at Greenwich
Apparent time 15. 7. 22 at taking the Dist. ☉ or ✶ & ☽
Difference ... 4. 10. 42 in time = p. 6 62. 40. 30
is Longitude between the Place of Observation and Greenwich.

N.B. Distance clear'd of Refraction { less 90. take the Diff.ce of the two Arches / More the Sum of the two Arches } is Principal Effect of Parallax { Arch first greatest - contra + / - from the Dist. clear'd of Refr.n }
By the Requisite Tables find the Parallax in Alt.de & by Table 4.th with Distance & Parallax in { Alt.de / Dist. } Difference sec.d corr.n Parallax.
which is to be + if distance is less 90. but more -
By the requisite Tables page 3. 4 & 5 with App.t Altitude ☽ and Hor.l Parallax gives the Parallax in Altitude

By Robert Bishop.

Publish'd according to Act of Parl.t Decem.r 9.th 1768.

G. 298: 1/3

Fig. 19 – Printed lunar-distance form, published by Robert Bishop, 1768

As the *Nautical Almanac* and timekeepers became, to a greater or lesser extent, more widely available, there was increasing interest in facilitating their use. Getting the new techniques on board at least a few ships, to be used and demonstrated by individuals trained in astronomy and mathematics, was an important initiative. There were also ongoing attempts to develop simplified astronomical and calculating methods, with Maskelyne soliciting and assessing ideas on behalf of the Board of Longitude. One approach was to provide printed forms that guided the user through the calculations and required significantly less mathematical ability (Fig. 19). There was also a recommendation from the Board in 1768 that Masters of Navy ships be required to attend the Royal Naval Academy at Portsmouth or selected London teachers for free instruction in how to use the *Almanac* and Hadley's quadrant. It was suggested that their future postings should depend on being able to produce a certificate of attendance.

The Admiralty acted but it turned out 'that great difficulties attend the obliging the old Masters of the Navy to perfect themselves in the use of those *Almanacs*'. They did not take kindly to the idea of compulsory land-based training and only fourteen Masters seem to have undertaken it in the two years following the Board's suggestion. The Board recommended instead that the Admiralty 'give such directions as they shall see fit for dispensing with such part of their said orders as related to the old masters and confining the same in future

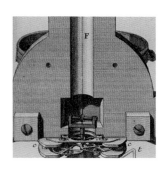

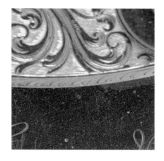

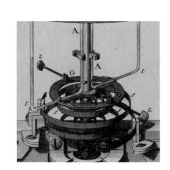

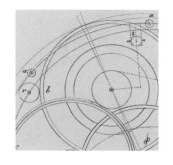
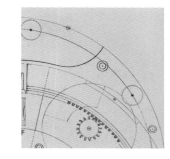

entirely to such new ones only as shall be henceforward appointed'.[28]

Despite Harrison's accusations, the Commissioners do not seem to have expended more energy on promoting astronomical rather than timekeeper methods, although it was understood that the former were more readily available and an essential complement to the early use of timekeepers. They spent large amounts of time and money on making, simplifying and trialling timekeepers. If they were not yet doing more to facilitate their use on board ships, that was because they were still very rare and expensive items.

Even with the efforts of key Commissioners of Longitude, and the involvement of King and Parliament, it is worth recalling how little impact the debates and decisions of the 1760s and 1770s had on the wider public. Little had yet changed in maritime practice. William Emerson, an eccentric mathematician and inventor who published on navigation, wrote in 1770 that as far as most were concerned the longitude was:

> still a secret, and likely to continue so. For tho' many thousand pounds have been paid for the pretended discovery thereof; I doubt we shall still remain just as wise as we were before the discovery; except the ill success of it happens to teach us so much wit, as to take better care of our money for the future.[29]

CHAPTER 5

WORKING AT SEA

We steered for St Helena in a straight line, trusting to our Watch and Lunar Observations.

John Elliott recalls the *Resolution's* homeward track on Cook's second voyage, 1775[1]

The Board of Longitude's work to support the lunar-distance and timekeeper methods was only a part of the story. It was also essential to show that the new techniques could be deployed in any ocean. This meant further trials, training sailors and making better charts. Marine timekeepers remained rare and expensive; it took decades to establish their reliability and for seafarers to trust them. Establishing the use of astronomical techniques, which required mathematical ability and application, was even more of a challenge.

One arena in which the new methods were trialled was during voyages of exploration. From the 1760s, the Royal Navy sent out a few special expeditions, most famously those of Captain James Cook (1728–79). These well-funded ventures were a testing ground for the new timekeepers and an opportunity for training mariners. The same was true of expeditions mounted by France, Spain, Russia and other seagoing powers, all international competitors with an equal interest in exploration and economic exploitation. In true Enlightenment spirit, these voyages gathered thousands of artefacts (Fig. 1) and reams of data, including geographical information from the new longitude techniques.

Fig. 1 – 'Various articles at Nootka Sound', by John Webber, showing items collected from the coast of what is now Vancouver Island on the north-west coast of North America during Cook's third voyage (1776–80)

Proof in the Pacific: the Cook voyages

Between 1768 and 1779, James Cook (Fig. 2) led three ambitious circumnavigations that would have a lasting impact on European visions of the Pacific. On the first (1768–71), he commanded the *Endeavour*, which was sent to Tahiti to make observations of a transit of Venus, with additional secret orders to search for a Southern Continent. The second (1772–75), with the *Resolution* and the *Adventure*, was to prove whether or not the supposed Southern Continent really existed, while on his third expedition (1776–80), which returned after Cook's death, the *Resolution* and the *Discovery* were to look for a Pacific opening of the long-sought North-West Passage. The voyages were central to the longitude story, since they allowed new instruments and techniques to be tested and applied to navigation and surveying.

Other than occasional privateering ventures, Britain had shown little interest in the Pacific until the outbreak of war with Spain in 1740, when George Anson was sent there with a small squadron, of which only his *Centurion* completed the full voyage (see Chapter 3, Figs 9–11). With the publication of Anson's narrative of his four-year circumnavigation in 1748, the challenges of navigating in unfamiliar waters became clear to all. Importantly, Anson's account stressed the need for proper surveys to produce accurate charts and prevent future voyages falling victim to the same problems.

Fig. 2 – Captain James Cook, by Nathaniel Dance, 1775–76. His right hand is pointing at the east coast of Australia on his own chart of the Southern Ocean

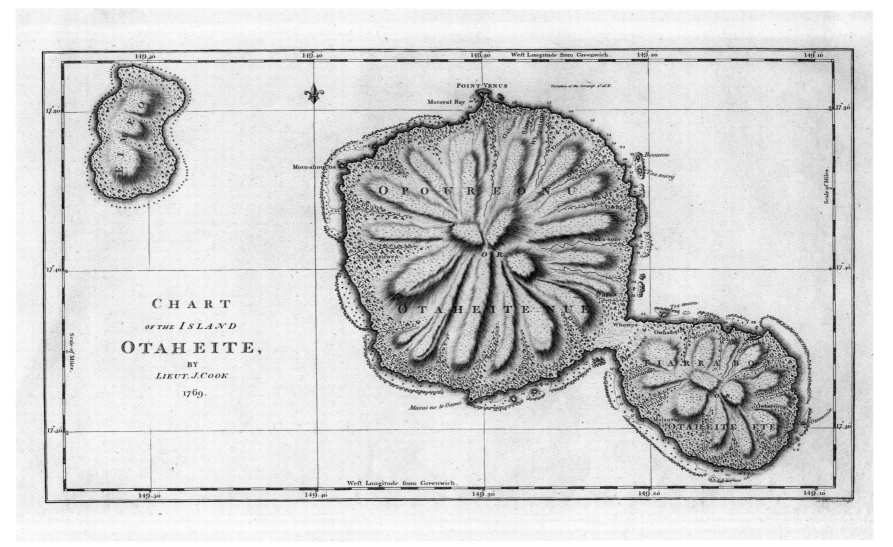

Fig. 3 – 'Chart of the Island of Otaheite [Tahiti]' by James Cook, 1769. He established his base for observing the transit of Venus at Point Venus

Anson's foray was followed up after the end of the Seven Years War with Royal Naval expeditions by the *Dolphin* in 1764–66 and again in 1766–68, accompanied by the *Swallow*. On the second of these voyages, the *Dolphin* discovered Tahiti (Fig. 3), and so determined where Cook's first expedition would go. This voyage was also one of the first occasions on which a British seaman made a reasonably accurate longitude determination in the Pacific, for the ship's purser (coincidentally named John Harrison) had the *British Mariner's Guide*, an octant and the skills to establish Tahiti's position. The commander of the ship, Samuel Wallis, said it was done by 'taking the Distance of the Sun from the Moon and working it according to Dr Masculine's Method which we did not understand'.[2] John Harrison was clearly an unusual purser.

Cook was sent to Tahiti because it was in the region that Nevil Maskelyne had suggested for observations of the next transit of Venus, due in 1769. From then on, the island became a regular port of call for British vessels and a fixed point at which navigators and astronomers could check their instruments. The Royal Society planned the expedition, with the King's support, and Maskelyne defined its astronomical purposes and equipment. The choice of Cook as leader was unexpected, but by this time he was an experienced navigator who had shown great skill in conducting surveys and had published solar eclipse observations in the *Philosophical Transactions*, the Society's journal. Another notable addition to the *Endeavour's* complement was Joseph Banks (1743–1820), who paid his own way as the expedition's naturalist. In later years, Banks would draw on his experiences with Cook as he became the guiding hand for all British voyages of exploration.

With the tools and tables developed before the *Endeavour* sailed in the summer of 1768, the expedition also offered the perfect opportunity to test the lunar-distance method in unknown waters. Astronomical expertise came from Charles Green, appointed by the Royal Society to assist with the observations at Tahiti. Noting as they left England that 'no person in the ship could either make an observation of the Moon or Calculate one when made', Green began to put this right.[3] Cook himself was using lunars by the time they reached Brazil. As they passed through the waters that had almost defeated Dampier's and Anson's ships, the captain had gained confidence in his new-found skills:

Fig. 4 – Cook's *Resolution* in the Marquesas Islands, by William Hodges, 1774

the Longitude of few places in the World are better ascertain'd than that of Strait Le Maire and Cape Horn being determined by several observations of the Sun and Moon, made both by my self and Mr Green the Astronomer.[4]

By the end of the voyage, the *Nautical Almanac's* value had also become obvious. With only the 1768 and 1769 editions available when the *Endeavour* sailed, from 1770 Cook and Green had to master the onerous calculations that the *Almanac's* tables normally took care of. This extra work aside, Cook was quick to emphasize that

> by [Green's] Instructions several of the Petty officers can make and Calculate these observations almost as well as himself: it is only by such means that this method of finding the Longitude at Sea can be put into universal practice – a method that we have generally found may be depended upon to within half a degree; which is a degree of accuracy more than Sufficient for all Nautical purposes.[5]

This is not to say that Cook depended on the new techniques. Rather, much of the navigation involved traditional methods, notably dead reckoning, and like other commanders before him he recognized the importance of local knowledge. Where possible, he enlisted the help of Polynesians, including Tupaia (*c*.1725–70), a priest and navigator originally from the island of Ra'iatea (in the western Society Islands). Tupaia joined the *Endeavour* at Tahiti as a local guide but stayed on board and aided Cook's negotiations in New Zealand, where the Maori welcomed him as an important religious leader.

Impressed with the evident success of Polynesian navigation, Cook attempted to learn more from Tupaia, having him draw a chart of the islands around Ra'iatea. Such efforts were part of an ongoing quest to document and understand the peoples they encountered, yet it is clear that Cook never fully understood practices that were not rooted in the coordinates of latitude and longitude, but which relied on a detailed knowledge of environmental features and patterns of change over a huge geographical area. Regrettably, Tupaia was one of those who died of fever later in the voyage.

Shortly after Cook's return to England in 1771, plans were afoot for a second voyage in the sloops *Resolution* (Fig. 4) and *Adventure*. This time the Board of Longitude was directing the astronomical aspects and appointed William Wales and William Bayly as observers. With new longitude timekeepers now ready, Maskelyne also suggested that the expedition

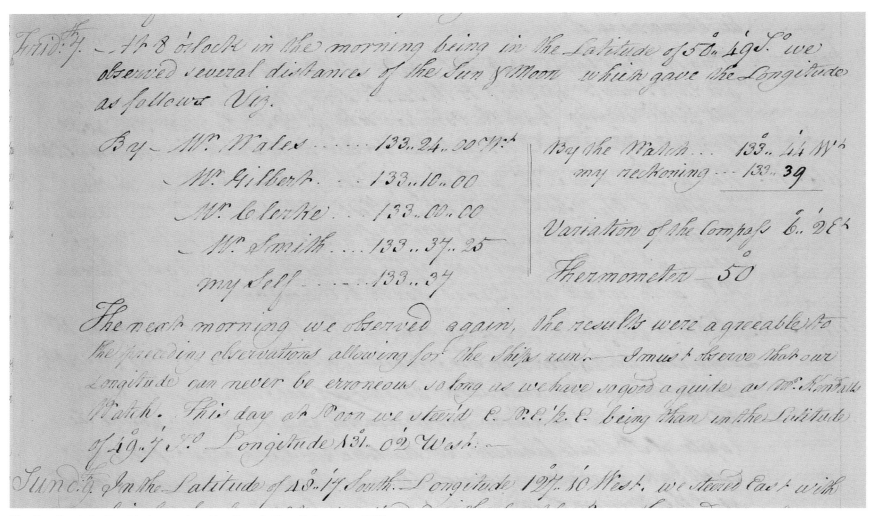

Fig. 5 – Captain Cook's journal on the *Resolution*, 7 January 1774, showing the longitude determined by lunar distance, timekeeper and dead reckoning (detail)

Fig. 6 – View of Point Venus and Matavai Bay, Tahiti, by William Hodges, 1773, probably sketched from the cabin of the *Resolution*

may be rendered more serviceable to the improvement of Geography & Navigation ... if the ship be furnished with such astronomical Instruments as this Board hath the disposal of or can obtain the use of from the Royal Society and also some of the Longitude Watches, and, above all, if a proper person could be sent out to make use of those Instruments & teach the Officers on board the ship the method of finding the Longitude ...[6]

The 'Longitude Watches' were Larcum Kendall's first marine timekeeper, K1 (Chapter 4, Fig. 11), and three by John Arnold (Chapter 4, Fig. 12), with their testing rigorously controlled much like the tests of H4 and H5. In this case, each watch was enclosed in a box with three locks, with keys held by the commander, first lieutenant and astronomer of the relevant ship, all three of whom were to be present for the daily winding and checking.[7] The procedure was generally followed but was not flawless. With Cook and his first lieutenant kept on shore in June 1774, Wales could not open the box and the watch ran down. However, as Cook wrote, 'This circumstance was of no consequence as Mr Wales had several al[t]itudes of the sun at this place before it went down and also got some after.'[8] The watch could be reset from the observations. As well as keeping the watches running, the astronomers had to monitor their performance, checked against dead reckoning, lunar-distance observations and other astronomical observations on land; in other words, they had to compare the results of all available methods for fixing positions (Fig. 5). The Board of Longitude issued instruments including sextants for this work and other equipment for investigations such as magnetic observations (see Fig. 9).

The timekeepers were very much at the experimental stage and their performance reflected this. Arnold's did not do well, perhaps not helped when Bayly forgot to wind one of them. Two stopped during the voyage and the rate of the third increased throughout. By contrast, Kendall's watch performed superbly, Cook coming to see it as his 'trusty friend' and 'never-failing guide'.[9] By the time they reached the Cape of Good Hope on the passage home in 1775, he was a convert. He wrote to the Secretary of the Admiralty that, 'Mr. Kendalls Watch has exceeded the expectations of its most Zealous advocate and by being now and then corrected by lunar observations has been our faithful guide through all the vicissitudes of climates.'[10]

Kendall's timekeeper was no longer on trial. Setting sail for St Helena, Cook put his trust in clockwork rather than opting for the usual practice of latitude sailing. 'Depending on the goodness of Mr. Kendalls Watch', he wrote, 'I resolved to try to make the island by a direct course, it did not deceive us and we made it accordingly on the 15th of May at Day-break'.[11] Cook's words were plain, but confirmed that an otherwise routine leg of the voyage had established the efficacy of a second method for determining longitude at sea.

Fig. 7 (following page) – 'View of Maitavie [Matavai] Bay', by William Hodges, 1776. The *Resolution* and *Adventure* are shown at anchor and a large tent that Cook erected to house his coopers, the guard and sick members of his crew can be seen on Point Venus

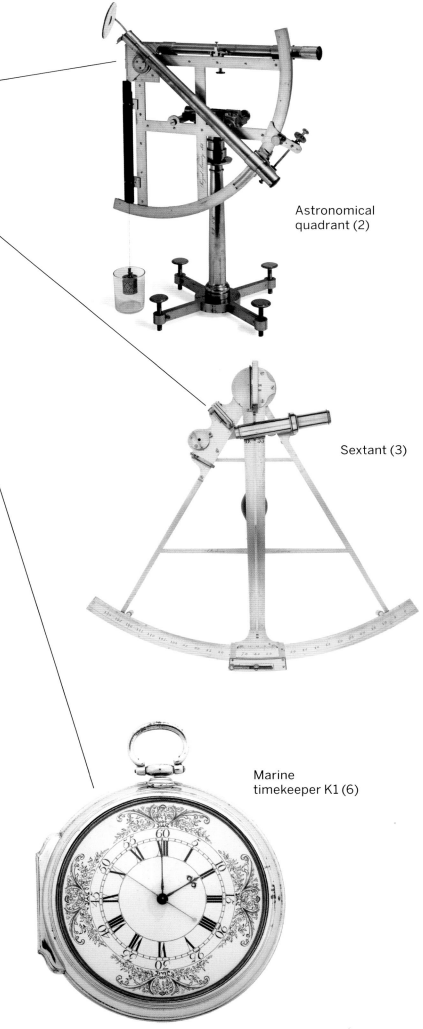

Astronomical quadrant (2)

Sextant (3)

Observing tent (7)

Marine timekeeper K1 (6)

on the coast of America and Asia greater care was taken to find the difference between Longitudes by Lunar obs^{ns} & by T.K. [timekeeper], these comparisons were made at Sandwich Sound, to the N of Cape Newenham & in Norton Sound, from whence I should suppose the ... Lunar observations on these coasts to be nearly the true situation of the ship in Longitude.[15]

In addition, lunar-distance and timekeeper methods were used both to check and supplement each other. During the second voyage, Cook found that they 'did not differ from each other two miles'.[16] On the third voyage, the longitude of Ship Cove in Nootka Sound was settled by 134 sets of lunars, ninety of them taken at the temporary observatory (Fig. 10), the remainder taken from the ship and corrected to the same position with the help of the timekeepers. The new, portable techniques could also fix the positions of places that would otherwise remain unknown: 'Even the situation of such islands as we past without touching at are by means of Mr Kendalls Watch determined with almost equal accuracy', Cook wrote in November 1774.[17]

Yet there was a hierarchy: astronomical determinations were considered more certain than those from mechanical timekeepers alone. In the later parts of the third voyage, the astronomical observations took on particular importance as the timekeepers

Fig. 10 – 'Resolution and Discovery in Ship Cove, Nootka Sound', by John Webber, 1778. Just to the right of the centre can be seen the astronomers' observing tents

began to cause concern. By 1779, K1's performance was 'so irregular that I fear very little dependence is to be put upon it'.[18] Finally it stopped and, though restarted, worked poorly for the rest of the voyage. The ships relied instead on Kendall's third timekeeper, K3, although it never performed as well as K1.

Astronomical methods were not perfect, though. Cook realized on the second voyage that, during the first, his and Green's observations had placed much of the South Island of New Zealand (Fig. 11) too far east. Likewise, on the third voyage, Lieutenant King worried about 'the extream errors in different peoples observing with different sextants', which might undermine confidence in the lunar-distance method.[19] He explained that the errors came from lack of practice and the swapping of sextants (Fig. 12) between the young officers.

King's concerns highlighted another important role for the voyages: training officers in longitude techniques. This had a longer-term legacy that was crucial for embedding astronomical and timekeeper methods within naval practice. From the second voyage onwards, the astronomers were to 'teach such of the Officers on board the sloop as may desire it the use of the Astronomical Instruments & the Method of finding the Longitude at sea from the Lunar Observations.'[20] Cook was optimistic about this: 'any man with proper applycation and a little practice' could learn the techniques if enough instruments and astronomical

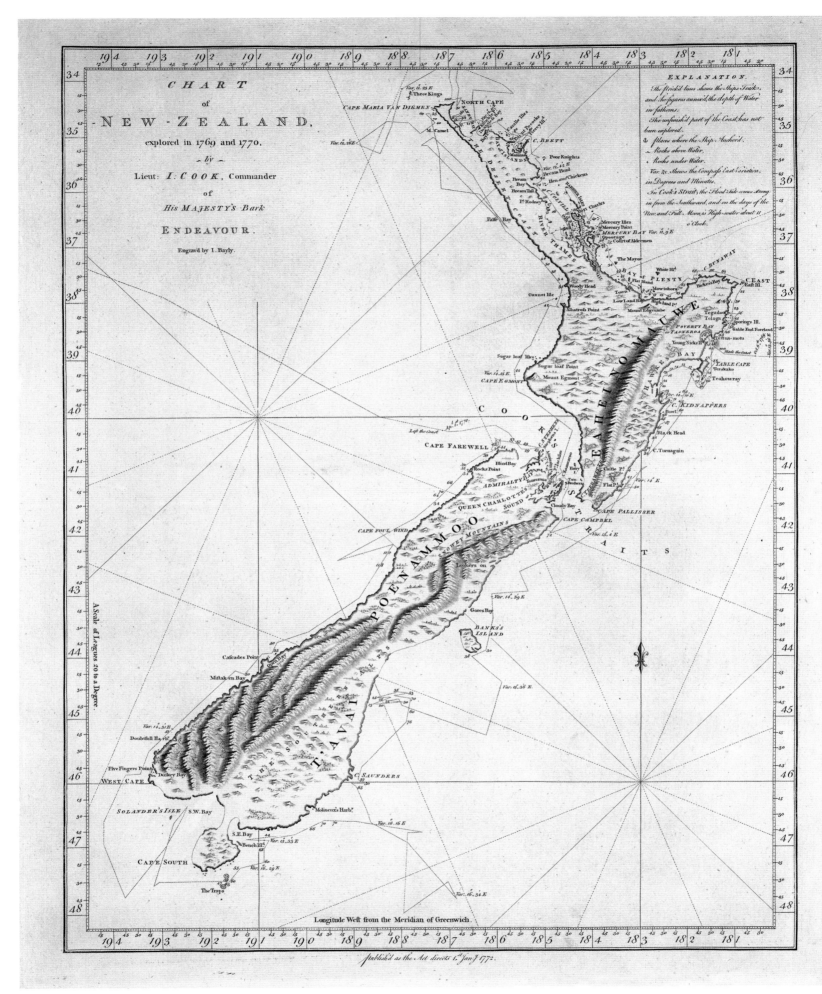

Fig. 11 – Chart of New Zealand, by James Cook, 1772. On the second voyage, Cook discovered errors in some of the longitudes determined during the first

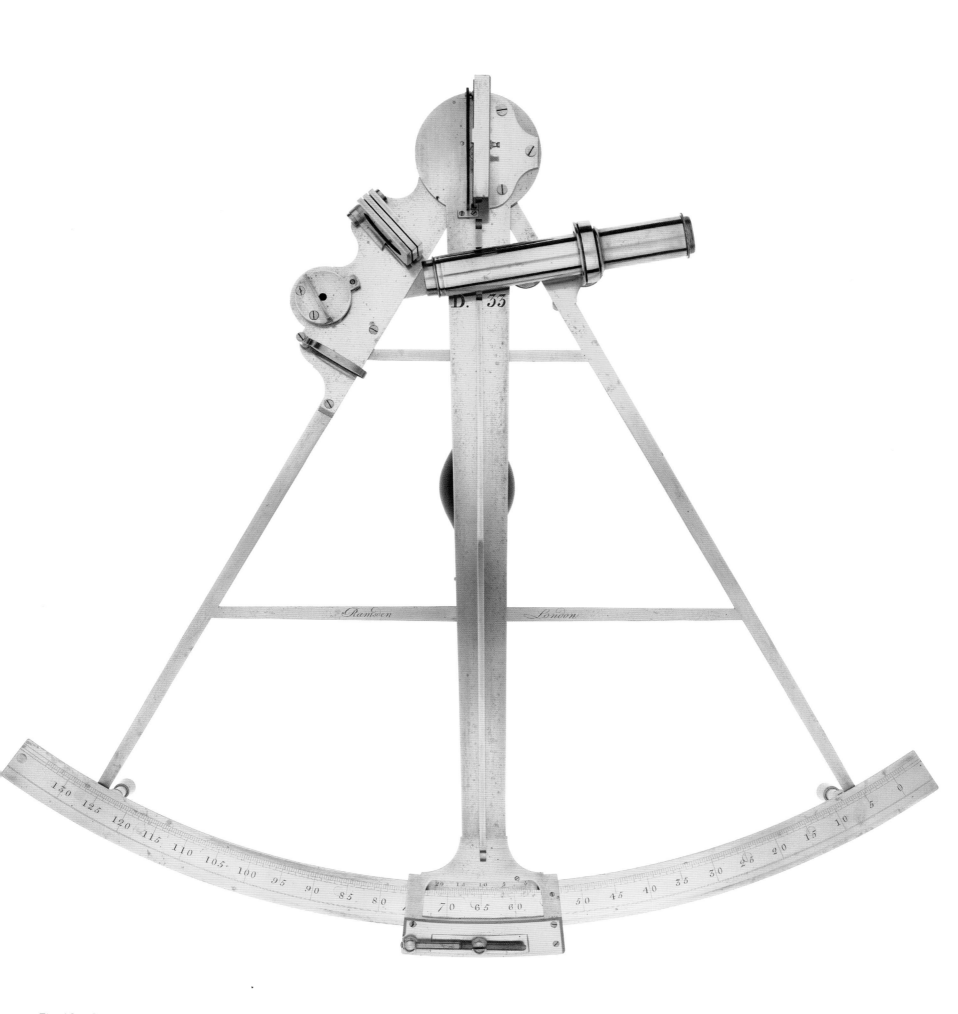

Fig. 12 – Sextant, by Ramsden, London, *c*.1772, used on Cook's third voyage

tables were supplied.[21] King agreed, although the abilities of these two mathematically adept officers probably blinded them to the difficulties others had with the calculations for lunar-distance and local time. This is borne out by an anecdote from 1781, when King was commanding a convoy to the West Indies. According to his first lieutenant, he chose to use lunars to make a direct course to Barbados, to the consternation of the ships being escorted:

> Nothing could exceed the surprise and terror of the masters of the merchant ships who, used only to the old jogtrot of their ancestors, were soon bewildered and lost all kind of tolerable accuracy in their reckoning. But when they found that the skill of the circumnavigators had brought them exactly to the desired point, nothing could exceed their admiration and astonishment.[22]

Cook, King and others showed that astronomical methods and timekeepers could be successfully deployed for navigation and surveying, and that the skills required could be passed on, but there was still some way to go before the methods would be commonplace.

In Cook's wake

James Cook's death during a skirmish at Kealakekua Bay in Hawaii on 14 February 1779 added to an already heroic reputation. To the leaders of later voyages, he became the archetypal explorer, although as Georg Forster, naturalist on the second voyage, wrote, 'there are no large discoveries left to make because now the globe is known from one end to the other.'[23] Nonetheless, the three voyages established the template for future expeditions by the inclusion of specialist astronomers, naturalists and artists. Later voyages also sought to exploit the discoveries Cook's expeditions had made of valuable products such as timber, flax, whales, seals and sea otters (Fig. 13), as well as new trading goods. Yet while the promise of lucrative trade drove many of the voyages, state-sponsored expeditions continued to lead the way in the development, testing and application of the new longitude methods at sea and on land.

Fig. 13 – 'A Sea Otter', by John Webber, artist on Cook's third voyage

The expense and rarity of the new instruments, particularly marine timekeepers, meant that they were available only to vessels that were well supported by the state or to individuals with cash to spare. As a result, specific instruments might serve several different voyages during their working life. Kendall's first marine timekeeper (K1), for instance, stayed in service beyond Cook's second and third voyages. Although it had stopped after Cook's death, Kendall was able to repair it and in 1786 the watch was passed to Captain Arthur Philip, sent out in the *Sirius* in command of the First Fleet to establish a colony in Australia. This was a very different type of voyage to Cook's but its status was high. The timekeeper was nearly lost when the *Sirius* was wrecked on Norfolk Island in March 1790 but, luckily, it was safely taken off before the ship sank. After returning to England, K1 was sent with John Jervis to the West Indies at the start of the French Revolutionary War, where it was presumably involved in operations, before being returned and withdrawn from service in 1802.

Kendall's second marine timekeeper, K2 (Fig. 14), had an even more turbulent history. Its first major naval expedition was on the *Racehorse*, commanded by Captain Constantine Phipps in a voyage towards the North Pole in 1773. Its most infamous years, however, came after it joined the *Bounty* in 1787. As Lieutenant William Bligh (1754–1817) later noted:

Fig. 15 – A branch of the breadfruit tree, from John Hawkesworth, *An Account of the Voyages Undertaken ... in the Southern Hemisphere* (London, 1773)

> The Object of all former voyages to the South Seas, undertaken by command of the present majesty, has been the advancement of science, and the increase of knowledge. This voyage may be reckoned the first, the intention of which has been to derive benefit from these distant discoveries.[24]

The aim of the voyage was to transplant breadfruit (Fig. 15) from Tahiti to the West Indies as food for the slaves forced to work the plantations there. This would benefit the plantation owners and thus Britain's expanding trade and imperial aspirations, and it was heavily promoted by Joseph Banks, a man of considerable influence after becoming President of the Royal Society in 1779. Banks understood the plant's potential from his time with Cook in Tahiti, where it grew in abundance. By 1787, he had convinced the government to support the expedition and then oversaw its planning, including the choice of ship and its conversion to accommodate breadfruit plants, as well as the selection of Bligh, a fine navigator who had sailed with Cook, as commander. Although the expedition was not primarily scientific, its status and Banks's sponsorship persuaded the Board of Longitude to lend Kendall's second timekeeper, a few other instruments and copies of the *Nautical Almanac*.

The voyage has since become notorious because of the events that culminated on 28 April 1789 with Fletcher Christian leading

Fig. 14 – Marine timekeeper K2, by Larcum Kendall, London, 1771

146

a mutiny and setting Bligh and eighteen others adrift in the ship's launch (Fig. 16). The mutineers recognized the value of K2 and other navigational paraphernalia. As the men being cast adrift collected supplies, Bligh recalled,

> Mr. Samuel [the Clerk] ... got a quadrant [octant] and compass into the boat; but was forbidden, on pain of death, to touch either map, ephemeris, book of astronomical observations, sextant, time-keeper, or any of my surveys or drawings.[25]

Bligh nevertheless managed to guide the twenty-three foot launch more than 3500 nautical miles to the Dutch settlement on Timor. Although he had neither charts nor timekeeper, he did have a sextant by Jesse Ramsden, an octant, a compass, a log line made from what was to hand in the launch and two navigation books, Hamilton Moore's *Practical Navigator* and Dunthorne and Maskelyne's *Tables Requisite*. This was enough for latitude observations and dead reckoning, although Bligh played up his navigational skills by later omitting to mention that he had the sextant and Hamilton Moore's book. Nonetheless, the boat's successful arrival in Timor was testament to the value of older techniques when deployed by a skilled navigator. Equally impressive was Bligh's ability to maintain the crew's morale and eke out their meagre rations.

Meanwhile, Kendall's watch remained on the *Bounty*, although whether it was much used is uncertain. Christian had some navigational knowledge, which presumably helped him choose their final destination of Pitcairn Island. This had been sighted 'like a great rock rising out of the sea' by Philip Carteret on the *Swallow* in the 1760s. Carteret believed that, though tricky, a landing would be possible and noted trees and running water, but his recorded longitude was wrong by three degrees.[26] As a result, even Cook was unable to find it. Christian must have realized that the island would make a safe haven from any ships sent after the mutineers. His thinking was sound: they arrived on Pitcairn in 1789 and K2 remained undisturbed until an American whaling captain discovered their refuge in 1806 and purchased it from John Adams, the last *Bounty* survivor. After passing through several hands and being stolen at least once, the watch arrived back in Britain in the 1840s.

The Board of Longitude supplied instruments to voyages of exploration because they wanted them tested, ideally under the supervision of observers they had employed. Two later voyages show how the Navy's own navigators also began to adapt to the challenges of deploying the new instruments and techniques at sea. Above all, they demonstrate increasing interest from the Navy in adopting promising longitude methods, in particular from officers who had been introduced to them during their time with Cook.

Fig. 16 – 'The Mutineers turning Lt Bligh and part of the officers and crew adrift', by Robert Dodd, 1790

Fig. 17 – Marine timekeeper K3, by Larcum Kendall, London, completed in 1774, showing the timekeeper, its mechanism and the timekeeper in its box. See page 154 for a detail of the mechanism

Fig. 18 (facing page) – 'A Chart showing part of the Coast of N.W. America', drawn by George Vancouver, published by Robinson and Edwards, 1798

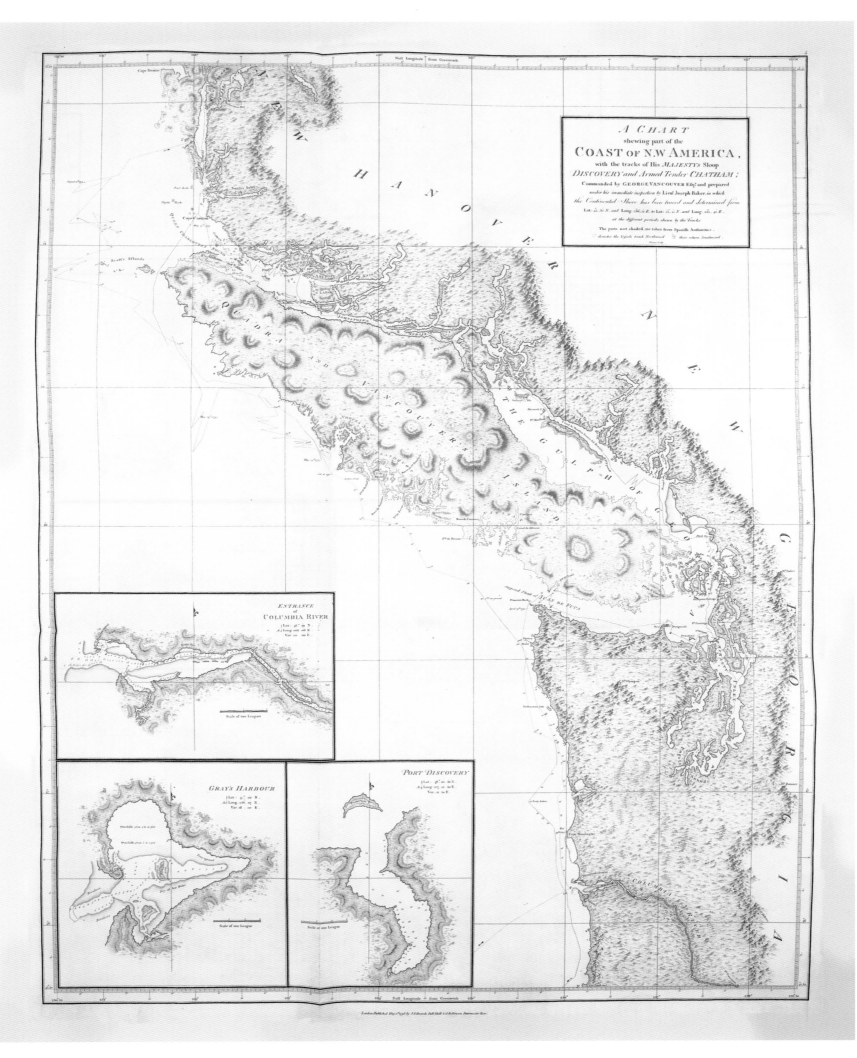

In 1791, Captain George Vancouver was sent to chart the north-west coast of America, to search for a North-West Passage and to resolve tensions with the Spanish following the recent Nootka Sound Crisis. It was a naval expedition, so it was the Navy Board that supplied the instruments, although these were similar to the equipment typically provided by the Board of Longitude. The Board did, however, supply two timekeepers: Kendall's third, K3 (Fig. 17), for the *Discovery* – a new ship of that name – and an Arnold for the *Chatham*. As a veteran of Cook's second and third voyages, Vancouver was familiar with longitude-finding techniques and could be trusted with the valuable equipment. Indeed, he was delighted with K3, 'the excellence of which', he wrote, 'had been manifested on board the [old] *Discovery* during Captain Cook's last voyage, and which had lately been cleaned and put into order by its very worthy and ingenious maker'.[27]

Vancouver was an enthusiast for lunar distances too and took advantage of the outward journey to test the equipment to hand. He tried out no less than twelve sextants and found that they 'agreed exceedingly well together'. Later, having observed 'many very good lunar distances of the sun and stars on different sides of the moon', he took their average to be the 'true longitude' and so judged the timekeeper's longitude to be 14′ 25″ too far east. Again, a hierarchy was clear in the perceptions of the different methods. Vancouver was keen to teach his junior officers as well, anticipating a time 'when every sea-fearing person capable of using a quadrant, will, on due instruction, be enabled by lunar observations to determine his longitude at sea.'[28] As he noted of the *Discovery*,

> on our departure from England, Mr. Whidbey [the ship's master] and myself could be considered as the only proficients in this branch of science; but now, amongst the officers and gentlemen of the quarter deck, there were several capable of ascertaining their situation in the ocean, with every degree of accuracy necessary for all the important purposes of navigation.[29]

The quality of the surveys and charts from the voyage (Fig. 18) showed that his trust in his instruments and newly trained crew was warranted. His findings that the rates of the timekeepers appeared to be affected by temperature variations also inspired many experiments after his return. Indeed, these same effects would be the subject of ongoing research throughout the nineteenth century.

The Board of Longitude's official instructions to their astronomers, and Vancouver's personal enthusiasm in passing on his knowledge, emphasized the importance of training for the diffusion of the new navigational techniques into the Navy. This was true in the case of Matthew Flinders (1774–1814), who initially taught himself navigation and astronomy with the encouragement of his father and a cousin. Having gone to sea in 1790, Flinders developed his skills under Bligh (on his second successful breadfruit voyage in the *Providence*) and Captain John Hunter, with whom he sailed to New South Wales in 1795. After his arrival there, he spent time exploring and surveying, which brought him to the attention of Joseph Banks, who was promoting a systematic coastal survey of New Holland, as Australia was then known. At Banks's instigation, a Royal Naval expedition was agreed, with Flinders given command of the sloop *Investigator*.

Flinders set off in 1801. As for Vancouver's voyage, the Board of Longitude lent its support and appointed John Crosley, a Greenwich-trained astronomer, who was given the usual instruments, including two box timekeepers made by Thomas Earnshaw (1749–1829, see Chapter 6), two by Arnold and sextants by Ramsden and Dollond. But Crosley left the ship at the Cape of Good Hope, complaining of 'Chronic Rheumettisa in my knees and ... Gaut in my foot'.[30] So the Board sent a replacement, James Inman, and supplied him with K3. With no Board of Longitude astronomer present in the meantime, the surveying and other work fell to Flinders and his brother Samuel, who had to learn to manage the timekeepers and other instruments. This was no easy matter. The observing tent was rotten, which made using the instruments difficult, and although the Earnshaw timekeepers performed quite well, two of the Arnolds stopped, one of them having made an 'uncommon noise'.[31] The new longitude methods could be successful but were fragile.

Inman finally arrived in June 1803, only to find that the *Investigator* had been declared unseaworthy. Keen to execute his astronomical duties, he made a series of observations in Sydney but found that K3 was behaving erratically. He blamed the variable weather. Meanwhile, Flinders decided to return to England in the *Porpoise*. Within a week he was shipwrecked off the Australian coast. Returning by boat to Port Jackson, he set off again in a small schooner whose leaky condition forced him into the French island of Mauritius. Unbeknown to him, war with France had resumed and he was consequently held as a prisoner there for six and a half years. Inman made a swifter return to England and on the way used the timekeepers to help fix the position of Wreck Reef, where the *Porpoise* had come to grief (Fig. 19).

Competition and cooperation

Oceanic discovery, especially in the relatively unknown Pacific, was never the domain of Britain alone. Every European seafaring nation was seeking to explore and exploit previously unknown parts of the world and, just as on British voyages, the expeditions of other nations began to test and deploy the latest navigational methods.

France's first major venture was led by Louis-Antoine, Comte de Bougainville (1729–1811), in 1766–69. Having suffered heavy losses in the Seven Years War, France was looking for new regions to colonize. Bougainville was ordered to search for these and expand trading opportunities, particularly with China. Predating the Cook voyages, it was also the first such expedition to take scientific specialists, including naturalists and an astronomer, Pierre-Antoine Véron (1736–70), who experimented with several new navigational techniques and instruments. These included the intriguingly named *mégamètre*, an alternative to the octant, and different astronomical methods for finding longitude. Tahiti's position, for example, was

> ascertained by eleven observations of the moon, according to the method of horary angles. M. Verron had made many others onshore, during four days and four nights, to determine the same longitude; but the paper on which he wrote them having been stolen, he has only kept the last observations, made the day before our departure. He believes their result exact enough, though their extremes differ among themselves 7° or 8°.[32]

This method used measurements of latitude and the Moon's altitude to determine its hour angle (its angle west of the observer's meridian) and, by comparison with a reference location such as Paris, the longitude. Véron had some success with it but Bougainville and others preferred lunar distances, aided by the French astronomical almanac, the *Connaissance des Temps*.

Among Bougainville's crew was Jean-François de Galaup, Comte de Lapérouse (1741–c.1788). Twenty years later, by then a well-respected officer, Lapérouse was chosen to lead another major expedition intended to regain French prestige once more. As was by now the norm, Lapérouse's ships, the *Boussole* and the *Astrolabe*, took specialists and the best instruments available, including five timekeepers by Ferdinand Berthoud, an English pocket timekeeper and at least four English sextants.

The voyage nevertheless ended badly. Having set off in August 1785, Lapérouse's ships spent time exploring and charting the Pacific. Early in 1788 they arrived in Botany Bay – just after Philip's arrival with the First Fleet – and departed again the following month, having left copies of Lapérouse's journals, charts and letters with Philip for transmission back to France. These last letters told how pleased Lapérouse was with the performance of the timekeepers but this was the last heard of the expedition. Its fatal end was deduced in 1826 and the ships' sunken remains were only discovered in 1964, on the reefs of Vanikoro, New Caledonia. There was just one survivor, Jean-Baptiste Barthélemy de Lesseps, who had disembarked in Kamchatka in September 1787 with other

Fig. 19 – 'Wreck Reef Bank', by William Westall, August 1803

Fig. 20 (following page) – Lapérouse's officers with the islanders on Sakhalin in the north Pacific, from *The Voyage of La Pérouse Round the World* (London, 1798)

DRESS of the Inhab

...nts of LANGLE BAY.

copies of logs, charts and letters, and trekked across Siberia for over a year to deliver them to Paris. These would be the basis of the first published voyage accounts (Fig. 20).

While Alejandro Malaspina's voyage of 1789–94 for the Spanish navy avoided the Pacific reefs, it too ended inauspiciously for its commander, who was arrested for conspiracy against the state just fourteen months after his return. He was imprisoned for six years and then exiled. The planning took place in happier times, however, as a response to Lapérouse's voyage, and was intended to create much needed charts of the farthest reaches of the Americas. It was also to be an imperial inspection of Spain's territories in South America and the Pacific.

The voyage's scientific aims allowed the Spanish to draw on the expertise of other European countries, Britain included. When it was realized that there were insufficient instruments in Cadiz, for example, steps were taken to purchase them in London. The hydrographer Alexander Dalrymple (1737–1808) was among those who advised on the equipment, which included astronomical instruments by Ramsden and Dollond, timekeepers by Arnold and the *Nautical Almanac*. The expedition also looked to France, purchasing two timekeepers by Berthoud and the *Connaissance des Temps*. Again, the expensive new technologies were precious. Malaspina's diary recorded that to keep one of the Arnold timekeepers safe, he had it 'hanging from my shoulder and held very close to my chest, so that there was no space for it to move, but it was rather cushioned by my own body'.[33]

Cook's influence on the Malaspina expedition was clear, right down to the name of one of the ships: *Descubierta* (Discovery). Like Cook's expedition, it included artists, naturalists, astronomers and other experts, with surveying as one of its key activities. If anything, the results were even more impressive than Cook's, comprising one of the largest collections of data ever assembled by a single expedition, including 450 notebooks of astronomical and hydrographic observations, 1500 hydrographic surveys, 183 charts and 361 coastal views. The technologies and methods of longitude determination had been fully applied in this colossal collection. Unfortunately, little of it was published at the time – the achievements of the expedition suffered from their association

see page 148, Fig. 17

with the outcast Malaspina, and were consigned to oblivion for more than a century.

It was not just at sea that the new techniques could be applied. In the areas of North America under the control of the Hudson's Bay Company, surveyors used sextants, artificial horizons and watches, together with the *Nautical Almanac* and *Tables Requisite*, to map the vast territory of Rupert's Land, then part of British North America. They had great success but encountered many hazards. In December 1791, the surveyor Peter Fidler had just completed his sextant observations when a scaffold loaded with meat fell on him, although 'by the greatest good luck the Instrument was unhurt.'[34]

The most famous land-based expedition of the period was by Meriwether Lewis and William Clark, who in 1804–06 crossed the United States of America to find a 'direct & practicable communication across this continent, for the purposes of commerce' and so link America's Atlantic and Pacific coasts.[35] Thomas Jefferson had spent twenty years considering this ambitious plan by the time he became President in 1801. Once in power, he pushed ahead a project that would symbolize the political and geographical aspirations of the young republic.

Jefferson was eager that Lewis and Clark deploy the latest instruments and techniques, and enlisted the help of the American Philosophical Society. Robert Patterson, professor of mathematics at the University of Pennsylvania, prepared material for field use, including forms for lunar distances, confident that his system for calculating longitude was 'easy even to boys or common sailors of moderate capacities'.[36] The expedition was also supplied with appropriate equipment, including an octant, a sextant, artificial horizons, a gold-cased chronometer by Arnold and the *Nautical Almanac*. Yet the observers' inexperience, compounded by the vagaries of the timekeeper, rendered the results disappointing. The best data came from dead reckoning and the more straightforward latitude observations. Nonetheless, the explorers returned with a remarkable collection of information about the people, flora, fauna and geography of the vast regions they had crossed, allowing the first accurate maps to be produced.

Although they formed just a fraction of maritime activity in the period, the voyages of exploration of the late eighteenth century showed that timekeeper and astronomical methods of longitude determination could be applied at sea. As the decades progressed, more people were trained and more instruments became available.

Yet it was a slow process and widespread adoption was not inevitable. Thomas Brisbane, astronomer and former Governor of New South Wales, recalled that on a naval expedition in 1795 'there were perhaps not ten individuals who could make a lunar observation'.[37] Timekeepers were gradually becoming more common too but remained rare until the nineteenth century. Small wonder, then, that Matthew Flinders could write about the new navigational techniques as curiosities in a whimsical biography of his seafaring cat, Trim:

> Trim took a fancy to nautical astronomy. When an officer took lunar or other observations, he would place himself by the Time-keeper, and consider the motion of the hands, and apparently the uses of the instrument, with much earnest attention; he would try to touch the second hand, listen to the ticking, and walk all around the piece to assure himself whether or no it might not be a living animal. And mewing to the young gentleman whose business it was to mark down the time, seemed to ask an explanation. When the officer had made his observation, the cry of Stop! roused Trim from his meditation; he cocked his tail, and running up the rigging near to the officer, mewed to know the meaning of all those proceedings.[38]

Maritime trading enterprises such as the East India Company also had an interest in the development of accurate navigational techniques, and their officers and surveyors were notably early adopters, but it was a piecemeal process. Dead reckoning remained the norm on merchant vessels worldwide until around the 1830s, when more and more began to use chronometers. On the occasions when lunars were used, it was as a check for positions established by dead reckoning or, later on, by chronometer.

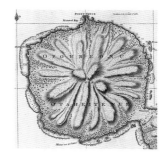

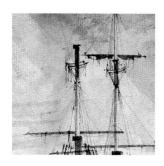

The seemingly slow take-up of new navigational techniques and equipment has often been put down to mariners' inherent resistance to change. However, it might be more appropriate to ask why anyone risking their life at sea should adopt a new technique or instrument without reasonable assurance of the benefit. It needed to be absolutely clear that the new methods offered tangible improvements and were easy to use, reliable and affordable. There also had to be a support system in place to supply and maintain the appropriate tools and to teach sailors how to use them. Both new methods for finding longitude could seem questionable by these criteria. Although the *Nautical Almanac* and tabulated forms speeded up the calculations for the lunar-distance method, laborious mathematics still faced the mariner wishing to use the heavenly bodies. The timekeeper method required similar calculations for determining local time and the watch's rate. This cannot have suited every aspiring navigator.

In the years before large-scale production of marine timekeepers, their reliability could not be taken for granted and some sceptics voiced their concerns about trusting them. William Wales had these doubters in mind when he railed against those whose efforts were 'bringing timekeepers into disrepute, and ... defeating the endeavours of the Board of Longitude, who have been labouring incessantly for the last 30 years to establish the use of them'.[39] Even after the East India Company put a column for longitude by chronometer in its log-books in 1791, captains began their daily reckoning not from this position but from the previous day's estimation by dead reckoning.

Yet there were successes. From the 1760s onwards, the new methods were exploited, albeit in a specialist capacity, for surveying and improving the charts on which navigation depended. Positive reports in published voyage accounts, and the gradual spread of officers who had learned the techniques on voyages of exploration, encouraged their use more routinely. Combined with support from the Board of Longitude, Royal Navy and East India Company, this would pave the way for formal and widespread adoption in the nineteenth century. For that to happen, however, there needed to be great changes in the production of timekeepers and observing instruments to make them more accurate, more reliable and, above all, more affordable.

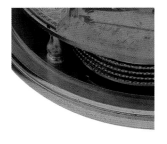

CHAPTER 6

COMMERCE AND CREATIVITY

Were we required to characterise this age of ours by any single epithet, we should be tempted to call it, not an Heroical, Devotional, Philosophical, or Moral Age, but, above all others, the Mechanical Age.

Thomas Carlyle, 'Signs of the Times', *Edinburgh Review*, 1829[1]

Between the 1770s and 1830s, changes in the manufacturing of the new instruments for measuring longitude made them cheaper and more widely available. These developments relied on many things: skilled artisans; well-managed systems for coordinating work; technical refinement; business acumen; and state encouragement, in part through the Board of Longitude. As markets developed, the canniest entrepreneurs were able to take advantage of the opportunities they offered. Competition was intense, however. Business rivalries could become bitter and attempts to profit from invention often came to nothing.

The wider transformation of Britain in this period, usually known as the Industrial Revolution, had a number of interlinked roots, including the introduction of new ways of using energy sources such as coal, harnessing steam and water power, and the beginnings of factory systems of production. These changes were not confined to Britain but had far-reaching effects there, fostered by relative political stability and an intellectual climate that encouraged the exchange of ideas through diverse publications, new organizations such as the Society for the Encouragement of Arts, Manufactures and Commerce, and a growth in public lectures and scientific demonstrations throughout the country. It was above all an age of innovation, as satirists reminded the public (Fig. 1). But innovation could be unwelcome. After the ending of the traditional procession to public hangings at Tyburn in 1783, Samuel Johnson complained that,

> The age is running mad after innovation; all the business of the world is to be done in a new way; men to be hanged in a new way; Tyburn itself is not safe from the fury of innovation.[2]

Fig. 1 – 'March of Intellect – Lord, how this world improves as we grow older', by William Heath, published by Thomas McLean, London, 1829

Although there were countless inventions, from the steam engine to the iron boat, their impact was normally felt only decades after their first appearance, once their teething problems were eliminated and they had been refined and made cost-effective. As Daniel Defoe had already noted in 1726, this was something at which British artisans and entrepreneurs excelled:

> most of our great Advances in Arts, in Trade, in Government and in almost all the great Things, we are now Masters of and in which we so much exceed all our Neighbouring Nations, are really founded upon the Inventions of others.[3]

These were skills that would be important in the development of instruments for determining longitude. At the same time, Britain's commercial landscape provided fertile ground in which ingenious and enterprising individuals could use the market to exploit their ideas. It was the age of the technological entrepreneur, made prosperous from the fruits of his own invention (see Figs 5, 7 and 11).

Mechanization and standardization were key features of the Industrial Revolution, but their spread was neither uniform nor rapid. Clock-, watch- and instrument making largely remained craft industries. In some other areas, too, attempts to displace humans with machines failed. Likewise, while standardized parts began to be developed for instruments, their introduction did not automatically lead to mass production, since the market for navigational equipment, though sizeable, was not so large that it was immediately essential.

Making the chronometer

John Harrison had shown that he could make a watch that kept good time at sea. As those who had seen him take H4 apart, and others who tried to make sense of the published account of it, realized, it was ingenious, extraordinarily complex and fiendishly difficult to copy. Timekeepers of equal accuracy were needed but had to be reproducible and affordable. Unfortunately, it was not immediately obvious how this might be achieved. One approach, unsuccessful as it turned out, was to build timekeepers based on Harrison's beautiful but complex machines. It was an approach taken by Thomas Mudge (c.1715–94, Fig. 2), one of the finest watchmakers of the period.

Fig. 2 – Thomas Mudge, by Nathaniel Dance, c.1772

Fig. 3 – Marine timekeeper 'Green', by Thomas Mudge, 1777

Fig. 4 – Mudge-type marine timekeeper no. 4, by Howells and Pennington, London, c.1794, in a later box

As a former apprentice of George Graham, Mudge had almost certainly met Harrison in the 1730s. Even if he had not, he was clearly impressed by Harrison's craftsmanship and by the early 1770s had turned his attention to the design of marine timekeepers, hoping that he could gain similar rewards from the Board of Longitude. He was, of course, already well known to the Board as one of the expert witnesses to the 'discovery' of H4, and as the man Ferdinand Berthoud had talked to about Harrison's secrets. His horological expertise was evident in his watches, which set the standard for quality in the second half of the eighteenth century.

Mudge's first marine timekeeper was completed in 1774, two years before Harrison's death. It tested well when sent to Thomas Hornsby, Professor of Astronomy at Oxford, yet stopped twice after being put on trial at Greenwich. Mudge blamed poor handling, so it was tried again from late 1776 to early 1778 and performed exceptionally at first, but then accelerated and stopped. Nonetheless, its performance convinced the Board of Longitude to award Mudge £500, although he was already unhappy that a new Longitude Act had limited the rewards on offer for timekeepers to £10,000, and introduced the requirement that two timekeepers be tested.

Undeterred, Mudge set about making two almost identical timekeepers, known by the colour of their cases as 'Blue' and 'Green' (Fig. 3). These were extraordinary pieces of mechanism, every bit as complex as Harrison's, but with a number of differences, notably in the escapement. The two were tested at Greenwich in the 1780s but did not perform well enough for the Board to approve either a reward or further trials. Mudge responded, through his son (a lawyer), by challenging the Board's decision and accusing Maskelyne and his staff of incompetence. Perhaps remembering the resolution of Harrison's dealings

Fig. 5 – John Arnold and family, by Robert Davy, c.1783

Fig. 6 – Pocket chronometer no. 36, by John Arnold, London, 1778

with the Board, the Mudges took the matter to Parliament and published a long account of the affair. Uncharacteristically, this led to a published rebuttal by Maskelyne and, less surprisingly, a counter-response by Mudge junior.

Meanwhile, a Select Committee of MPs heard evidence from all concerned, while members of the Board of Longitude voiced their opposition to Mudge's petition. Joseph Banks lobbied particularly hard in print and behind the scenes, writing personally to the Committee members. Rewarding Mudge, he argued, would be to reward 'an inferior artist, manifestly to the injury & discouragement of those who are superior to him in the Same Line'. If they were to rule in Mudge's favour, he added, the Board would be 'sorely humiliated'.[4] To his horror, the Select Committee did just that and awarded Mudge a further £2500, a testament to his son's successful lobbying.

In the light of Parliament's decision, Mudge junior decided to market the designs and set up a dedicated factory under two London watchmakers, William Howells and Robert Pennington. The new venture failed, however, and, by 1796, Howells had formed a rival partnership. This failed as well; Mudge's designs were just too complex. In the end, the factory produced only twenty-seven timekeepers (Fig. 4), and Howells's breakaway firm no more than about seven.

With the benefit of hindsight, one can see the wisdom in the Board of Longitude's emphasis on modifying and simplifying Harrison's work to create a design that could be reproduced in large numbers. This was largely achieved through the work of two watchmakers: John Arnold, whose timekeepers were being trialled on Cook's second voyage, and Thomas Earnshaw. By combining essential elements from Harrison's designs and the concept of the detached escapement, almost certainly derived from the work of Pierre Le Roy in France, they were responsible for creating the marine chronometer as we know it.

Arnold and Earnshaw were very different characters. Arnold, as his family portrait suggests (Fig. 5), was a man with aspirations, who knew how to develop and exploit social and commercial networks. His manner was not to everyone's taste. Thomas Mouat, a Shetlander who met him in 1775, found his talk 'animated, blustering and much adorned with oaths'.[5] Nonetheless, Arnold's easy self-promotion was already clear in 1764 when he presented a miniature watch to George III, thus securing royal patronage that would help him build a lucrative business. Alongside productive dealings with the Board of Longitude, Arnold got to know Alexander Dalrymple and through him gained access to the East India Company. With Dalrymple's encouragement, Company captains purchased Arnold's timekeepers and commented enthusiastically on their performance.

He was an innovator too, whose commercial success arose in part from his technical improvements. In 1775, he patented the helical (or coil) balance spring, which was more isochronous than a flat spiral spring, and the compensation balance, which adjusted the shape of the balance to accommodate changes in temperature. The balance was incorporated into his watch no. 36 (Fig. 6), which performed so well in trials at the Royal Observatory that Arnold proudly published the results in 1780. He also noted that the name 'chronometer' had been suggested for the watch by Dalrymple and Banks. Dalrymple was also using the term in his own publications on the use of the new marine timekeepers in navigation and charting. This marked the adoption of the term in its modern sense. The watch's results impressed many others too, including the mathematician William Ludlam, who confessed his changed opinion of Arnold:

> I once thought he was one of those who talk much and do little ... But since the account of the going of his watch ... and since I have seen what he has actually done, I have far different thoughts. His contrivances are far simpler and easier than those of Mr Harrison's, or any I have seen, and full as likely as any to answer the end proposed.[6]

In 1782, Arnold took out a patent for a spring-detent escapement, a new type of detached escapement, which required no problematic lubrication. It was this patent that would cause a bitter and long-running dispute with the younger watchmaker Thomas Earnshaw (Fig. 7). Earnshaw was more of a maverick, blunt in his public statements, quick of temper and prone to accusation. As a businessman, he was less skilful than Arnold, already having spent

Fig. 7 – Thomas Earnshaw, by Martin Archer Schee, c.1808

Fig. 8 – Marine chronometer, by John Arnold, London, c.1784

Fig. 9 – Marine chronometers nos. 512 (top) and 524 (bottom), by Thomas Earnshaw, London, c.1800

Fig. 10 – Escapement model, made by Thomas Earnshaw for the Board of Longitude, 1804

Fig. 11 – Jesse Ramsden, by Robert Home, c.1791

time in the Fleet debtors' prison by the mid-1770s, but he was the finer craftsman and quickly gained a reputation for quality among London's watchmakers.

Earnshaw seems to have turned his attention to marine timekeepers around 1780. In contrast to Arnold's ever-changing approach to their design (compare Fig. 12 of Chapter 4 and Fig. 8 here), Earnshaw quickly settled on an arrangement that he stuck to and which would become standard for all marine chronometers (Fig. 9). He had simplified Harrison's complex designs to such an extent that his timekeepers had just 128 parts (plus the box). Like Arnold's, they included a spring-detent escapement, which Earnshaw said he had invented a year before his rival patented the device (Earnshaw did take out a patent, but not until 1783). Earnshaw had boasted of his invention to other watchmakers and maintained that Arnold must have learned about the new escapement from them. It was an accusation Arnold never denied, lending plausibility to Earnshaw's grievance, which he pursued aggressively for the rest of his life, even after Arnold's death in 1799.

In the meantime, gaining an introduction to Nevil Maskelyne in 1789 improved Earnshaw's fortunes. His work impressed the Astronomer Royal, who became a key supporter in later years. By the early 1790s, Earnshaw was in regular contact with the Board of Longitude, sending timekeepers for trial at the Royal Observatory and supplying them to numerous ships and expeditions. Their quality was often noted. Returning from the Cape in 1803, for example, John Crosley reported that the squadron's commander was using an Earnshaw timekeeper, which was 'the only timekeeper in the fleet to be depended upon'.[7]

Never one for the soft approach, Earnshaw was soon petitioning the Board for rewards, often in startlingly blunt terms. Eventually this paid off: in 1804 the Board decided to grant him £3000 if he would reveal his secrets, but they also decided that Arnold's years of work should be equally rewarded. This enraged Earnshaw, still embittered at Arnold's theft of his invention. So began a protracted dispute between the Board, Earnshaw and Arnold's son, John Roger (the boy in Fig. 5), who had taken over his late father's business. It was a contest that would rapidly embroil London's watchmaking community and expose divisions within the Board of Longitude, notably between Joseph Banks in support of Arnold and Nevil Maskelyne fighting Earnshaw's corner.

To resolve matters, the Board proposed that Earnshaw and Arnold junior produce models of the disputed escapements (Fig. 10), as well as written descriptions of their respective mechanisms, which were printed and circulated to other watchmakers for comment. The Board also interviewed London's watchmakers about the competing claims of Arnold and Earnshaw and about the manufacture of chronometers more generally, seeking to understand the trade and assess how near they were to their goal of having reliable chronometers produced in large numbers.

Earnshaw did himself few favours. His explanation, which was unfortunately printed without being vetted, was littered with further accusations against Arnold and with more generally scurrilous comments. 'Mr Arnold was a pompous man', he wrote, 'and because he made large Machines it must be right, and the fools followed him in his Blunders.' Arnold's machines were of an 'absurd size', Earnshaw went on, because they needed a large, powerful mainspring 'to drag the Works on through all Impediments of oil and dirt'.[8] His own, of course, were free of such flaws.

The book had to be pulped and a sanitized version reissued. Nonetheless, Earnshaw received only lukewarm support from his peers and the Board duly decided that both watchmakers should receive the rewards originally suggested. Earnshaw, still furious, then made matters worse by publishing a broadsheet and newspaper articles that, among other things, accused the 'malicious and envious Watchmakers', with Joseph Banks at their head, of 'extraordinary exertions' against him.[9] Banks considered Earnshaw's attack libellous enough to sue but the Board, perhaps with Maskelyne's encouragement, refused to take up the case. Unsatisfied with this lacklustre response, Banks threatened to pursue it personally, but eventually let it drop. Unsurprisingly, later petitions from Earnshaw failed to gain the Board's support.

Fig. 12 – Ramsden's second dividing engine, from *Description of an Engine for Dividing Mathematical Instruments* (London, 1777)

Banks, meanwhile, attended no further Board meetings until after Maskelyne's death.

Earnshaw's more positive legacy was the adoption of a procedure for the large-scale production of chronometers that endured until after the First World War. Arnold and others in the eighteenth century appear to have thought of their marine timekeepers as small clocks, each one a unique piece, and had all the parts made in London. Earnshaw, by contrast, thought of his as large watches, and followed long-established watchmaking practices – standardized rough movements were made in Lancashire and then sent to London, where the mainspring, dial and other parts were added, and the chronometers were finished, adjusted and cased. Finally, the retailer's name and number were added just before sale. The whole process typically involved more than forty specialist artisans but produced work of the highest quality.

Earnshaw's method, though large scale, still followed a pre-industrial system in which work was subcontracted to individual artisans, with central control lying in the standardization of the design. This was not the same as true mass production, as was being used for the manufacture of other, simpler items, such as buttons and snuffboxes. Traditional techniques were more suited to the existing market for marine chronometers, and Arnold and Earnshaw made more than 2000 box and pocket chronometers between them. This met more than half of Britain's naval and merchant shipping needs up to the 1820s. By comparison, French production in the same period was in the low hundreds. An 1819 encyclopedia article on chronometers could therefore reasonably hope that 'the spirit of competition for public fame will continue to entitle our English manufacturers to that preference among naval officers, which the excellence of their workmanship entitles them to expect.'[10]

Engines and scales

Changes were afoot in the manufacture of observing instruments as well. As early users of octants and sextants came to appreciate, hand-held instruments were limited by the accuracy of their degree scales and their size, both of which were determined by production methods. Until the 1770s, scales were

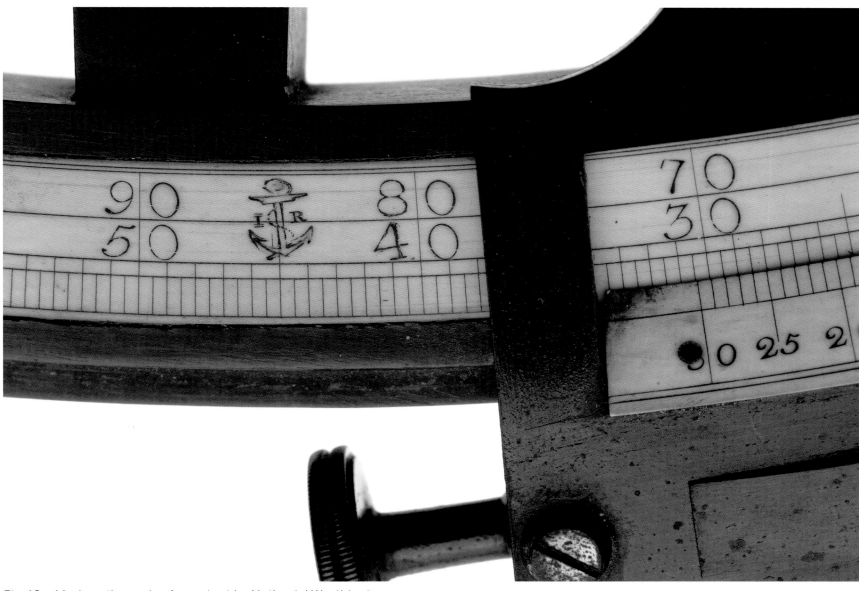

Fig. 13 – Mark on the scale of a sextant by Nathaniel Worthington, London, c.1840; the scale was presumably divided on the second engine made by Jesse Ramsden

Fig. 14 – 'Little Midshipman' trade sign, late eighteenth century

divided by hand using a beam compass, an engraving instrument for marking large circles. This was a highly skilled and laborious art, in which makers like George Graham and John Bird were acknowledged experts. Indeed, Bird's skills were so valued that the Board of Longitude paid him to describe his methods in print. The question of size was directly related, since the smaller the scale, the harder it was to divide accurately by hand. Observing instruments of the mid-eighteenth century tended, therefore, to be fairly large (see Chapter 3, Fig. 26).

This changed from the 1770s with the introduction of a mechanical scale division instrument known as a dividing engine. Its development was thanks above all to Jesse Ramsden (1735–1800), one of London's leading instrument makers, who had begun trying to devise a dividing engine in the 1760s. He produced his first in 1768, although this proved unsatisfactory, and six years later he completed a successful second engine; he sits proudly next to it in his portrait by Robert Home (Fig. 11). The first one was then sold to Jean-Baptiste Gaspard Bochart de Saron, President of the Paris Parliament, who had to hide it in some furniture to transport it secretly between the two countries, which were then at war. It was subsequently confiscated during the French Revolution and passed to the instrument maker Étienne Lenoir, who converted it to the Republic's metric scale (with 400 divisions of the circle instead of the usual 360 degrees) and used it to train younger craftsmen.

The second dividing engine consisted of a horizontal wheel with 2160 teeth incised around its rim (Fig. 12). These engaged on a precision screw, which was turned to rotate the wheel through predetermined angles. The instrument was clamped onto the wheel and, by alternately depressing the treadle and moving a cutter, the operator could divide its scale in thirty minutes with as much accuracy as a master divider could have done by hand in as many hours. Most importantly, the engine could graduate scales down to eight inches in radius, allowing smaller, more easily manageable instruments to be made.

The Board of Longitude recognized the value of mechanical division and published Ramsden's description in 1777, rewarding him for the invention itself and for turning it over to public use. Developing the approach it had taken with John Harrison, the Board increasingly felt that its role included offering rewards to makers if they disclosed their production methods, passed them on to other artisans and relinquished all rights to the public. As part of the deal with Ramsden, therefore, he had to teach up to ten instrument makers to build their own engines, and divide octant and sextant scales for other makers on his own engine (for a fee). These scales can often be identified from a special mark – in Ramsden's case, an anchor with his initials either side (Fig. 13).

It was not long before a few other makers had their own engines, some copying, others altering Ramsden's design. They too put their own marks on the scales they divided as a guarantee of their quality. By 1789, Jesper Bidstrup, a Danish instrument maker and industrial spy, was reporting that:

Fig. 15 – Models of (top) a scoring machine and (bottom) a mortising machine, by Marc Isambard Brunel and Henry Maudslay, c.1803. The full-sized machines were used to mechanize the production of wooden pulley blocks at Portsmouth dockyard

The division of instruments here is no longer free hand, unless their radius is 2 feet or more; everything is done on machines ... The owners of these machines will not permit anyone to see these machines, fearing lest others should get similar machines, by which they might lose their share of the advantage they have by dividing.[11]

The introduction of the dividing engine revolutionized the production of octants and sextants, making them smaller, cheaper and more widely available: Ramsden's firm, for example, had made at least 1450 sextants by the end of the century and divided countless scales for others. It was not long before these instruments were sufficiently common among mariners to become emblematic of the naval officer. A midshipman holding his sextant or octant (Fig. 14) was an instantly recognizable shop sign by the nineteenth century, with Charles Dickens in *Dombey and Sons* in 1848 describing these 'little timber midshipmen in obsolete naval uniforms, eternally employed outside the shop-doors of nautical instrument-makers in taking observations of the hackney coaches.'[12]

In contrast to chronometer makers like Arnold and Earnshaw, Ramsden brought his workers under one roof at his Piccadilly factory in London. This allowed him to control the production of the instruments that bore his name more closely than would be possible with the usual network of artisans working in their own workshops or houses. Ramsden's set-up was unique in instrument manufacture for this period and can be seen as part of the gradual move towards factory-style production, although his workers were still specialized in the same ways as those working in their own premises. Nonetheless, it was a sign of things to come.

The limits of machines

Mechanization was neither inevitable nor straightforward. A less successful attempt to mechanize the production of the tools of astronomical navigation came in the work of the Cambridge mathematician and scientific reformer Charles Babbage (1791–1871). 'One of the most singular advantages we derive from machinery', Babbage wrote, 'is in the check which it affords against the inattention, the idleness, or the knavery of human agents.'[13] With this in mind, he sought to improve the *Nautical Almanac* and other published tables by replacing human computers with mechanical ones.

Babbage had in fact applied to be a computer for the Royal Observatory after leaving Cambridge in 1814, but his friend John Herschel dissuaded him from what he was sure would be a thankless task. Later, on coming to appreciate the number of errors in the existing tables, Babbage is said to have exclaimed that, 'I wish to God these calculations had been executed by steam!'[14] Around the same time, the *Nautical Almanac* was coming under attack from many quarters, including the astronomer and actuary Francis Baily, who bemoaned the errors he claimed had crept in since Nevil Maskelyne's death. So it was that, in the summer of 1822, Babbage launched a project to build a machine, which he called a 'Difference Engine', that could automatically perform logarithmic calculations and print sets of tables. In a stroke, he claimed, it would eliminate errors and reduce losses at sea. For, as Herschel later wrote, 'An undetected error in a logarithmic table is like a sunken rock at sea yet undiscovered, upon which it is impossible to say what wrecks may have taken place.'[15]

Babbage's vision was of a machine that could perform complex calculations by reducing them to a sequence of additions, carried out by linked series of mechanical gears. It was an idea that emulated the latest innovations in automated manufacture, exemplified by the block-making machines introduced at Portsmouth dockyard (Fig. 15). Devised by Henry Maudslay and Marc Isambard Brunel (father of Isambard Kingdom Brunel) and installed between 1802 and 1807, the Portsmouth machines made the wooden pulley-blocks that the Navy needed in vast numbers – 100,000 a year by 1800. Operating as a production line, they could turn out 1420 blocks a day, allowing ten men to do as much as 110 by hand alone. This meant that fewer and less-skilled workers were needed, which led to a reduction in wages for those remaining. Resistance was inevitable but in the end futile.

As an early example of mass production in action, Maudslay and Brunel's machines exemplified precisely the sort of automated production line Babbage had in mind for his Difference Engine. It was no coincidence, therefore, that it was they who introduced Babbage to Joseph Clement, a skilled toolmaker and draughtsman, who would oversee the project. Its scale was breathtaking. The Difference Engine would need around 25,000 high-precision parts and weigh many tons but, with the completion of a small experimental version in 1822, Babbage was able to get the Royal Society's support and financial backing from the British government. Production of a full-sized engine began in earnest two years later. But the work took years and, with costs soaring, was brought to a standstill by financial disputes. Clement had nevertheless successfully assembled a working demonstration piece (Fig. 16), which Babbage proudly showed to all and sundry. Invited to witness the machine's workings after church one day, the American academician George Ticknor recalled that 'during an explanation which lasted between two and three hours, given by himself with great spirit, the wonder at its incomprehensible powers grew upon us every moment.'[16]

The halting of production was a major blow, although Babbage continued to seek government and public support, as well as devising new and improved designs including his 'Analytical Engine', a programmable computing machine that has led to him being called the father of the modern computer. Yet, despite his impassioned lobbying, experts were uncertain about the possibility and benefits of producing mathematical tables mechanically, particularly given that the government had already spent almost £17,500 on Babbage's incomplete machine by 1834. When consulted by the Treasury, George Airy (1801–92), Astronomer Royal since 1835, was in no doubt: Babbage's engine

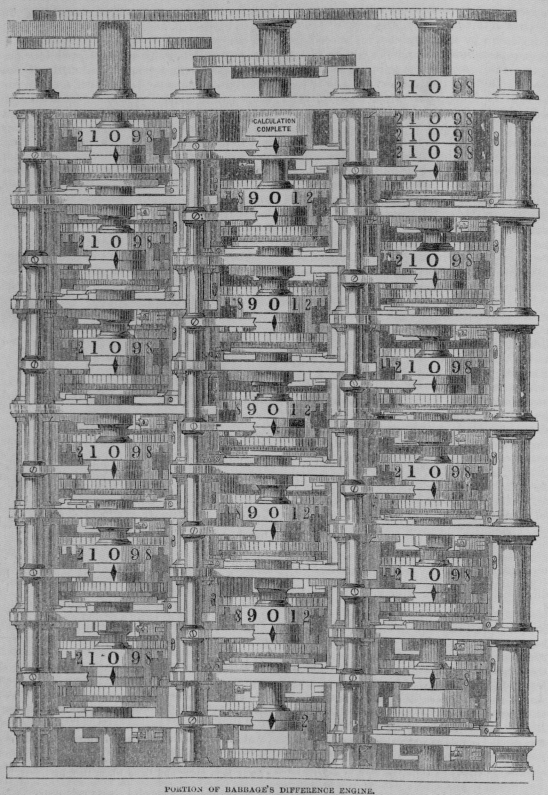

Fig. 16 – The working demonstration piece of Babbage's Difference Engine, from *Harper's New Monthly Magazine*, 1865

Fig. 17 – Four of the twenty-one volumes of Babbage's *Specimen of Logarithmic Tables* (London, 1831)

was 'useless'.[17] State backing was killed off in 1842 during a period of financial uncertainty when the government was looking to make substantial cuts.

Babbage was also a connoisseur of mathematical tables and had a personal collection of 300 or more volumes. He appreciated that a further source of error, in addition to calculation and printing errors, lay in users misreading the tables. Just one look at the printed masses of numbers is enough for one to appreciate how easy it would be to copy the wrong figure while in the middle of a lengthy calculation on a pitching and rolling ship. Babbage therefore looked into ways of making printed tables easier to read. Seeking out every colour of paper and ink available, he had sample tables printed in all combinations to see which would be the most effective (Fig. 17). He could not be faulted for his thoroughness: he even tried illegible pairings such as black on black. In the end these experiments also came to nothing.

The quest goes on

While the 1760s had seen the development of the lunar-distance and timekeeper methods for finding longitude at sea, the search for other methods and for advances in existing ones did not end. New legislation in 1774 extended the Board of Longitude's remit to include more general improvements in navigation, making them 'the Scientific Protectors of the British Navigation, as Castor and Pollux were of the old Ships of Greece and Rome'.[18] Other opportunities for would-be inventors also began to appear. From 1755, the recently founded Society for the Encouragement of Arts, Manufactures and Commerce offered premiums in mechanics, in particular for practical applications such as navigation, with a number of proposals for instruments and timekeepers being rewarded. Patenting offered a further means of trying to turn a profit from an idea. The marketplace did not want for new inventions – a few succeeded but many failed.

The Board of Longitude now found itself assessing increasing numbers of applications, which by the end of the century had to be sifted ahead of Board meetings. Joseph Banks warned one correspondent that the Commissioners would only respond to ideas they thought worth pursuing. Finally they decided that no proposal would be considered unless it had a supporting certificate from someone with authority. Many applicants got no response.

Most proposals sought to improve or extend the three existing longitude methods: dead reckoning, lunar distances and timekeepers. Among the ideas for dead reckoning, which was still used on every ship, the Board considered many for improved speed measurement and depth sounding. As one correspondent pointed out, dead reckoning was essential around the British coast, given the preponderance of 'cloudy weather, accompanied by dense fogs'.[19] His mechanical speed log failed to impress the Board but others were considered more seriously. Edward Massey, for instance, had a lengthy correspondence about his mechanical log, which became a commercial success (Fig. 18). The Board also awarded him £200 for a depth-sounding machine and suggested the Navy buy 500 of them.

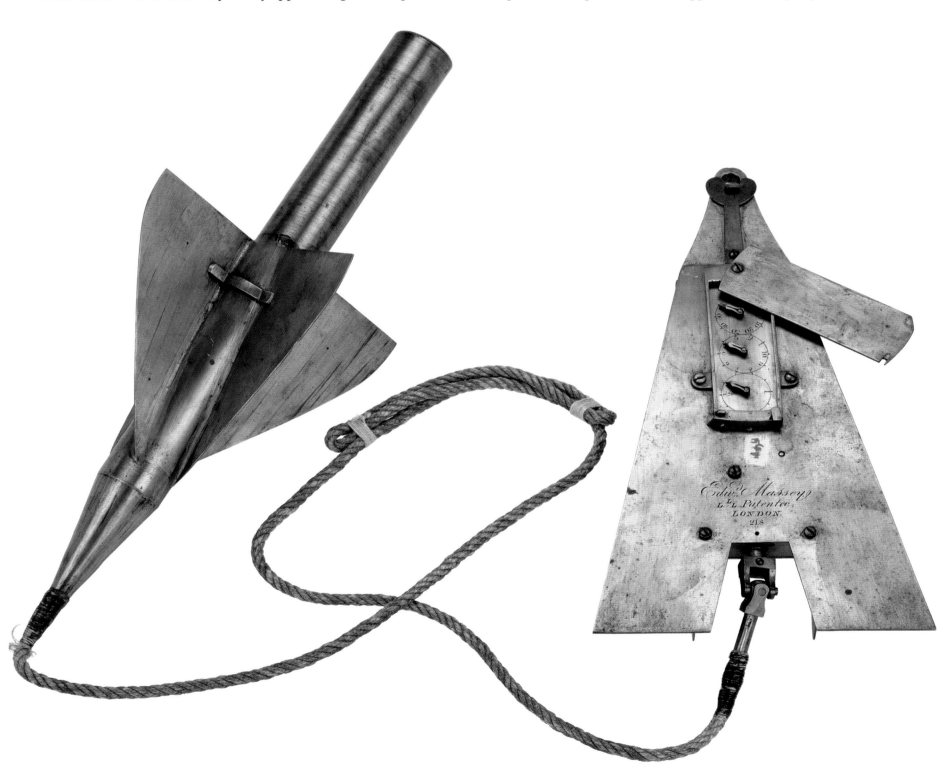

Fig. 18 – Mechanical log, by Edward Massey, London, c.1830

Fig. 19 – William Chavasse's proposed observing platform, submitted in 1813

179

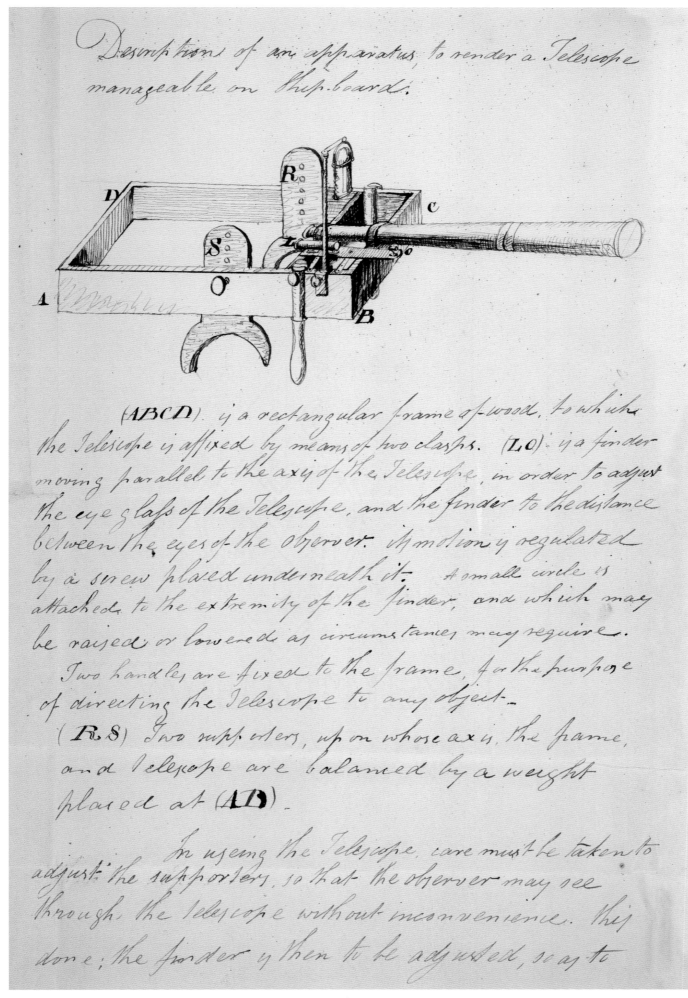

Fig. 20 – Samuel Parlour's shoulder-mounted apparatus for observing Jupiter's satellites, submitted in 1824

Other proposals looked to the ideas that were still proving impossible at sea, including observation of Jupiter's satellites. Although the failure of Irwin's marine chair had led Nevil Maskelyne to warn that managing a telescope on a ship was likely to remain a pipe dream, many applicants explored the idea. Joseph Senhouse of Arkleby Hall near Cockermouth wrote that he had trialled a model chair on a voyage to China by placing a wine glass filled with water on it. Not a drop was spilled, he said, and his full-sized chair worked perfectly. In 1813, Lieutenant William Chavasse of the 6th Madras Regiment sent a beautiful watercolour illustration (Fig. 19) of an observing platform but his scheme was not thought worthy of the Board's attention. Others imagined devices that sat on the body, like Samuel Parlour's shoulder-mounted arrangement (Fig. 20). Parlour had already tested it and the Board was interested enough to arrange further sea trials, but it proved unwieldy in strong winds.

The Board also considered proposals about magnetic variation. One of the more successful came from the engineer Ralph Walker (1749–1824, Fig. 21), a Scotsman who had settled as a planter in Jamaica. His compass (Fig. 22) had a sundial attachment to find true north from the Sun. Comparing this with the needle's direction gave the magnetic variation and thus, in theory, longitude. Walker had already impressed the Governor of Jamaica and Captain Bligh but Maskelyne was not convinced: it was neither innovative nor useful, he said. Nevertheless, the Board met Walker and sent the compass for trials. In the end, it was not felt to be a viable longitude solution but Walker received

Fig. 21 – Ralph Walker, published by James Asperne, 1803 (detail)

Fig. 22 – Azimuth compass, designed by Ralph Walker, London, c.1793

Fig. 23 – Mercury log glass, designed by Henry Constantine Jennings, made by William and Thomas Gilbert, London, c.1817

183

£200 for his improved compass design. The Navy also bought several, although some officers found them tricky to use.

Walker dealt with the Board well. Not so Henry Constantine Jennings, a chemist and inventor who, among other things, campaigned against the waste of stationery in the House of Commons. Jennings suggested several devices, including a mercury-filled log glass for measuring speed (Fig. 23) and an 'insulating compass' (Fig. 24). This, he said, always pointed due north, allowing mariners to determine latitude and longitude. Its secret was the addition of curved pieces of iron beneath the card. These were covered with specially prepared iron filings to protect the needle from magnetic deviation (deflections caused by iron in the ship), with more iron filings lining the compass bowl.

Although the Board of Longitude was sceptical, Jennings got the Admiralty and the East India Company to test his compasses. Captain John Ross took one on his 1818 expedition in search of a North-West Passage and found that it 'answered the purpose for which it was intended, and completely obviated the effect of local attraction; but ... ceased to act when the variation was great'.[20] Jennings took this as an endorsement and later claimed the support of no less than 2711 mariners. Sadly, he lacked diplomacy, blustering in one letter that, 'it requires only *Common Sense* to judge of the case; & I am sorry the Board of Admiralty have proved themselves so deficient in that necessary quality'.[21] His failure to gain any lasting support comes as little surprise.

Fig. 24 – Insulating compass, by Jennings and Company, London, c.1818

Any scientist today would dismiss Jennings's ideas as nonsense but in the early nineteenth century they had some plausibility. The same could not be said for every letter to the Board of Longitude. As the only government-funded body with a scientific remit able to give out money for new ideas, the Board found that it was presented with an extraordinary array of miscellaneous schemes. Lieutenant John Couch sent a stream of ideas between 1818 and 1823. Some concerned longitude methods, including the improvement of lunar tables. Others were more general: better ways of communicating between ship and shore; replacing the hay for cattle with dried parsnips and carrots to prevent shipboard fires; and a 'Calitsa' for carrying troops through surf (Fig. 25).[22]

Some proposals were more esoteric. John Bradley wrote with an impenetrable account of how he had found out the 'londdetude', offering to present it in person if the Commissioners would send a 'small Bill' to pay his way from Birmingham to London, 'for I am not able to walk so far'.[23] In 1819–20 the Board considered letters, diagrams and tables from Henry Croaker, beginning with his discovery of longitude from the 'perpetual motion' of the shadow on a sundial.[24] The Board did not respond. The more astute at least tried to link their ideas to finding longitude; perpetual motion machines, for example, might keep perfect time on a moving ship. The Board disagreed: perpetual motion lay far beyond its interests, as did squaring the circle and other impossibilities. Yet, as one correspondent observed, they were in danger of becoming 'the Grand National Tribunal for such difficult and obstruse undertakings'.[25]

When the Board of Longitude was finally disbanded in 1828, the political rhetoric played up the extent of these 'wild ravings of madmen, who fancied they had discovered perpetual motion and such like chimeras',[26] although the reasons for its demise had more to do with government cuts and Admiralty moves to create a smaller advisory body on scientific matters. In truth, most of the ideas the Board had been assessing were legitimate attempts to solve the problems of position finding at sea. Far from being over, the quest for longitude lived on. How it was to be encouraged and judged, however, was changing.

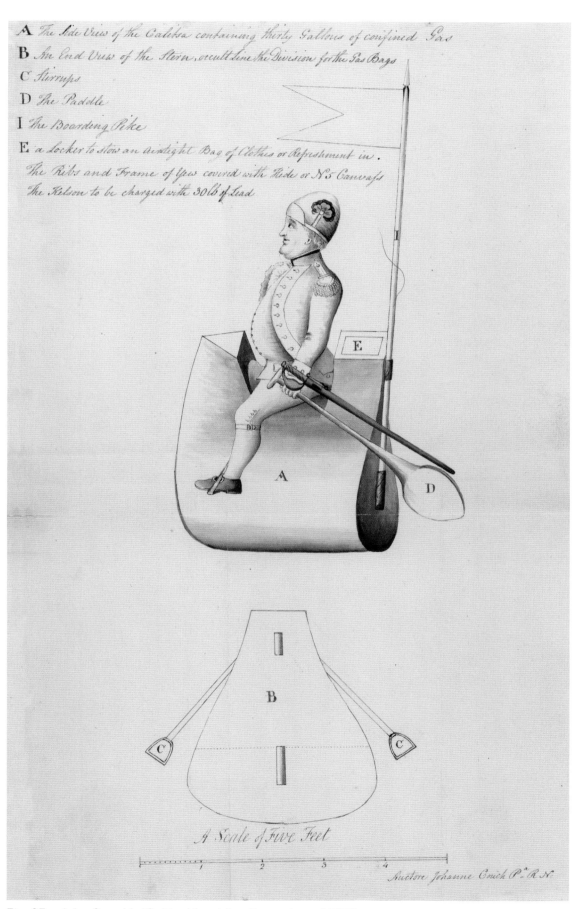

Fig. 25 – John Couch's 'Calitsa' for riding through surf, 1819

Britain's pride in its manufacturing expertise was in the ascendant in the last decades of the eighteenth century and would soon be justified in British global economic domination. Advances in the development of instruments for determining longitude were a conspicuous part of this, with British makers acknowledged as world leaders. The new devices could even become vehicles of imperial diplomacy, as happened during British attempts to woo the Chinese court in the hope of developing trade.

It was well known that the Chinese Emperor was fascinated by scientific instruments, so when Charles Cathcart led a mission for the East India Company in 1787, he took £488 worth of instruments (over a quarter of the value of all the gifts carried), including chronometers valued at £175. Cathcart died en route and the mission was aborted. It was revived five years later under George, Lord Macartney, with additional funding from the British government. Macartney's lavish gifts were mostly scientific equipment, including astronomical, surveying and navigational instruments

by Ramsden and Dollond, and a chronometer by Josiah Emery. Yet the mission failed. The Chinese saw the gifts as tribute from a lesser nation, luxurious novelties akin to the automata and instruments already common in court circles.

Nevertheless, there was every reason for pride in the skills of artisans who had made evident advances in the design and manufacture of horological and navigational instruments for finding longitude. This progress was bound up with broader changes in British industries, although the introduction of fully mechanized, factory-style processes remained a long way off. By the beginning of the nineteenth century, standardized instruments, and the texts and mathematical tables to go with them, were being produced in numbers large enough for their increasingly widespread use in naval and merchant vessels the world over. This did not end the quest for new or improved techniques but it did pave the way for the regular use of more certain navigational techniques, based on the longitude methods first recognized as practicable half a century earlier.

CHAPTER 7

Defining The World

In a country like ours, the commercial relations of which are so extensive, and whose ships, conveying the lives and property of our countrymen to distant climes, are daily trusting for safety and guidance across the pathless waters, to the researches of the astronomer, the maintenance of such an establishment is of the highest importance.

'The Royal Observatory, Greenwich', 1833[1]

By the end of the eighteenth century, the numbers of available timekeepers, sextants and astronomical tables, and of officers able to use them, were slowly increasing. However, this growth in the use of new instruments did not in itself revolutionize the safety of shipping. The records do not reveal a sudden drop in the number of wrecks, or even a reduction in maritime insurance premiums. Great dangers remained, with wrecks still caused by uncharted rocks – the fate that befell the *Magnificent* near the French port of Brest in 1804 (Fig. 1) – or bad weather. If there was a revolution, it was in the way that the longitude methods slowly affected the infrastructure and training associated with the Navy and merchant fleets.

The new instruments and techniques really came into their own on elite voyages of survey and exploration that followed in the wake of Cook, Vancouver and Flinders, and the lasting legacy of lunar distances, timekeepers and accurate instruments was in their application to the creation of reliable charts. It was this that ultimately improved safety and allowed increased world trade, and it was enabled by the Admiralty's limited but increasing support of surveys, specialist officers and technical publishing. In an age of naval peace, after the ending of the Napoleonic Wars in 1815, scientific skill could, for a handful, be a route to recognition and perhaps promotion.

While Britain was by no means the only nation engaged in such activity, it was Britain's dominance of maritime trade and empire in the nineteenth century that drove its effort to map the world's coastlines. The century opened with the publication of the first Admiralty Chart by the Hydrographic Office, which had been founded in 1795 to review and improve chart provision. By the 1850s they offered nearly 2000, with some 64,000 copies issued to the Fleet and more on sale to the public in Britain and overseas. Some of these surveys have, even today, not been superseded. Use of the *Nautical Almanac* in surveying as well as navigation meant that charts were based on data produced at the Royal Observatory, making the Greenwich meridian an increasingly significant reference point for the world.

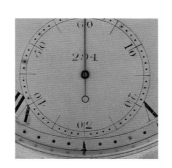

The very general diffusion of Chronometers of late years, and the great improvements that have recently been made in their construction, renders the management of them a subject of considerable moment.

Richard Owen, 'Essay on Chronometers', 1827[2]

Embedding the techniques

As well as remaining open to ideas for new navigational techniques, the Board of Longitude had a long-term role in supporting and improving existing methods. The *Nautical Almanac* continued to be produced, initially supervised by Nevil Maskelyne's successor as Astronomer Royal, John Pond, and later by the Board's Secretary, Thomas Young (1773–1829). From 1818, the Board became responsible for the Navy's stock of chronometers. This task was delegated to Pond in 1821, making the testing and rating of chronometers a core activity of the Royal Observatory for the next century and a half. As well as helping to make the use of chronometers reliable and routine, the Greenwich chronometer work encouraged improvement and innovation. Annual competitive chronometer trials were begun in 1823, with rewards and honour on offer to successful makers.

The East India Company, too, was routinely using chronometers. In fact, they adopted them, and complementary astronomical methods, considerably ahead of the Navy, reflecting the more difficult navigational challenges their vessels faced compared with most naval voyages. Time was money, so use of chronometers became normal practice, and by 1810 nearly every ship owned or hired by the Company had at least one chronometer. However, precedence was officially given to the use of astronomical methods and, in practice, to dead reckoning.

By 1821, 130 chronometers were listed as Admiralty property, of which forty-three were on voyages of exploration or surveys and fifty-eight on the seventy-four vessels then employed abroad (the rest being in storage). Seven ships had two chronometers, meaning that twenty-three of these overseas vessels, which since 1800 had been officially obliged to carry one, went without. Dead reckoning and, to a lesser extent on non-specialist ships, astronomical methods remained essential for many decades to come. Only in 1859 was it decided that ships commanded by captains should carry three chronometers – a number that allowed proper checks on performance – although other vessels still might have only one or two.

Checking, testing, trialling and issuing chronometers were not the only ways in which the timekeeper method could be supported. As Captain Robert Wauchope (1788–1862) had first suggested in 1818, what was also needed was a way of checking their rate, other than by unreliable on-board astronomical observations. He proposed that 'time balls' be put up in prominent places that would, by being raised and dropped at the same time each day, provide an accurate time signal to docked shipping. It took more than ten years and further lobbying before the first experimental example was erected in Portsmouth harbour. It was controlled by a visual signal from the Royal Naval College there, which had its own small observatory for teaching purposes and chronometer rating.

In June 1833, Wauchope suggested that a time ball be erected at Greenwich, visible to shipping in the Thames. This time the Admiralty acted swiftly and one was installed at the Royal Observatory later that year (Fig. 2). John Barrow (1764–1848), Secondary Secretary to the Admiralty and Commissioner of Longitude, issued a notice in the *Nautical Magazine*, explaining that the ball would drop daily at one o'clock: 'By observing the

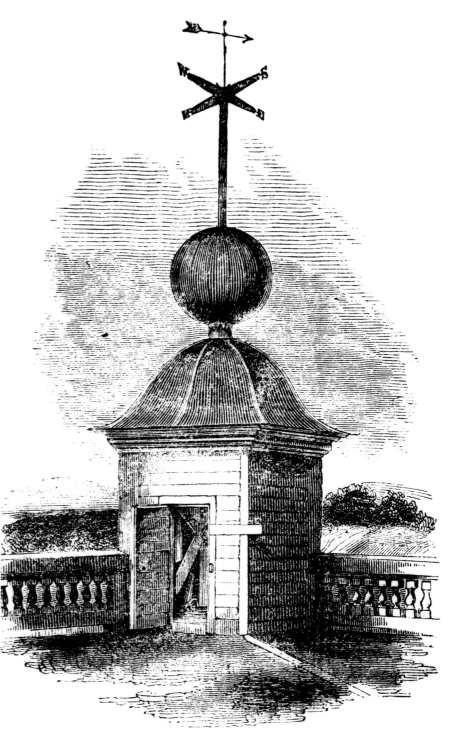

Fig. 1 (previous page) – 'Loss of the *Magnificent*, 25 March 1804', by John Christian Schetky, 1839

Fig. 2 – The time ball at the Royal Observatory, Greenwich, *Illustrated London News*, 9 November 1844

first instant of its downward movement, all vessels in the adjacent reaches of the river as well as in most of the docks, will thereby have an opportunity of regulating and rating their chronometers'[3]. The time ball still drops at 1 p.m. every day, though now more for the sake of tourists rather than for shipping.

Wauchope also persuaded the East India Company of the usefulness of this approach and, in the 1830s and 1840s, time balls came into operation at Mauritius, St Helena, the Cape of Good Hope, Madras (now Chennai) and Bombay (now Mumbai). Even if there was no time signal, observatories in docks and port cities provided a service to mariners. Their transit observations could define local time precisely, their longitude determinations allowed a conversion to Greenwich time, and many provided chronometer-rating services.

Elsewhere, efforts to improve the education of officers continued. Even in the early eighteenth century, developments in navigational techniques were making the process of learning seamanship significantly more complex. The arrival of precision observing instruments, chronometers and astronomical tables as part of normal equipment on board ship simply meant additional things to learn and, while observations were often a matter of learning by doing (Fig. 3), the theory and calculations had to be taught and practised.

Despite the existence of institutions like the Royal Naval Academy and the Royal Mathematical School, most midshipmen and officers learned navigational techniques at sea, perhaps supplemented by classes offered by the many teachers of navigation advertising in London and elsewhere. In theory, though not always in practice, there were also schoolmasters on board naval ships providing the education of future officers. From 1731, the 'Duties of the Naval Schoolmaster' were 'to employ his Time on board in instructing the Voluntiers in Writing, Arithmetick, and the Study of Navigation, and in whatsoever may contribute to render them Artists in that Science'.[4] In the first decades of the nineteenth century, there was a series of small improvements in the pay, conditions and status of naval schoolmasters, before they were replaced in 1837 with the higher-ranked naval instructors.

Whether or not there was an official teacher present, learning took place and the steps for finding position using the new techniques slowly became routine. Officers passed on their knowledge and there were many manuals of navigation. Increasingly, young midshipmen and others would practise their calculations and mathematical rules on board ship (Fig. 4), or when they had time ashore. One beautiful and painstaking example of a navigational workbook, from 1810,

Fig. 3 – Deck scene, with two men taking Sun observations from the quarterdeck, by Thomas Streatfeild, 1820

Fig. 4 – 'Life on the ocean, representing the usual occupations of the young officers in the steerage of a British frigate at sea', by Augustus Earle, c.1820–37. In the foreground one of the young men works on his mathematical calculations, with a sextant lying behind him. An octant and telescope hang on the beam above

was produced while its author was held prisoner (Fig. 5). Despite his circumstances, he worked through the problems outlined in two standard navigation textbooks, by John Robertson and John Hamilton Moore, that formed the basis of many more regular courses of education.

There were some attempts to improve official training on land, although it was by no means obligatory, and was not even the usual way to gain entry into the Navy or to achieve promotion. The Royal Naval College in Portsmouth was reformed in 1808 from the existing Naval Academy there, although its main function was the training of cadets. The college tried to hire good teaching staff and the curriculum focused on mathematics and the classics. Responding to the perceived need of the times, mathematics came first, with almost thirty hours of instruction and additional evening work devoted to it each week. A mathematical prize was instituted in 1819, the first winner being Robert FitzRoy (1805–65), who was an exemplary student and went on to make his career in hydrographic surveying.

In general, though, the College was not well regarded. It was closed in 1837, and it was decided that all elementary teaching would take place at sea, emphasizing the importance of practical experience and the judgement and patronage of existing officers. The College reopened in 1839 to provide 'additional means of scientific education to the young gentlemen of the fleet' – that is, specialist training for a few officers.[5] However, it quickly became a means of cramming for the lieutenant's exam, rather than a place of higher education, and there was some suspicion from captains who had not themselves received a land-based scientific training. They, apparently, 'looked upon scientific attainments not so much as a waste of time, but as injurious to the acquisition of seamanship and the details of routine'.[6]

It was not until the introduction of training ships in the 1850s and educational reforms in the 1870s that naval education became more universal and systematic. In the early part of the century, only a minority succeeded in gaining a higher mathematical and

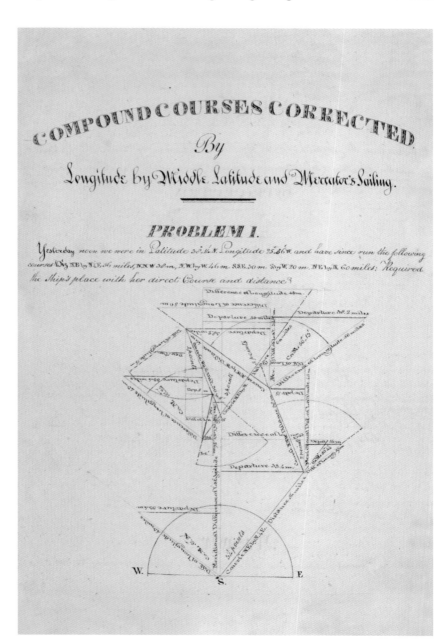

Fig. 5 – Pages from a navigational workbook compiled by John Marshall, 1810 while he was being held as a prisoner

navigational education, whether through what was essentially an apprenticeship system, the College or other means, such as the limited navigational stream of the Greenwich Hospital. Some of them were able to profit from their education by doing specialist work on the increasing number of naval survey expeditions, perhaps giving them an edge in an era of massive underemployment of qualified officers. Nevertheless, promotion still most often relied on skill and bravery in action, which after 1815 might be seen in, for example, the capture of a vessel engaged in illegal trade. Only a very few had an opportunity to make their name on the high-status expeditions that took their inspiration from Cook's eighteenth-century voyages.

The Board, Science and the Arctic

Having demonstrated, to its own satisfaction at least, the advantages of a scientific committee advising the Admiralty, the Board of Longitude expanded its remit in other directions too. It had committees to explore technical issues, such as how to develop better optical glass for observing instruments – work which was passed on to the young Michael Faraday at the Royal Institution – and how to establish a ship's tonnage correctly. An ambitious reworking of the Board and its role as an intermediary between the scientific world and government was masterminded by Joseph Banks and his close associates within the Admiralty, the Secretaries John Wilson Croker and John Barrow, who had, after Maskelyne's death, increasingly come to dominate proceedings.

The reworked Board was created by an 1818 Act of Parliament. This spelled out the ongoing rewards for improvements in navigation and the Board's role in overseeing chronometers and the *Almanac*. It also provided paid positions for Thomas Young and six scientific advisers, three of whom would be picked, essentially by Banks, from among the Fellows of the Royal Society. This Act also, somewhat surprisingly, revived a half-forgotten parliamentary reward designed to encourage exploration of the putative North-West Passage. This reward copied the 1714 Act by specifying different sums payable for varying degrees of success but otherwise attempted to tie the Board into a much broader maritime project.

The need for the continued existence of the Board was explained to Parliament by Croker. He declared that, although there were now two ways of finding longitude at sea 'with partial success, the one mechanical and the other scientific, it was essential to the ultimate perfection of the discovery, that scientific principles should predominate' and that much work was yet to be done.[7] While perhaps attempting to blind his fellow MPs with science, Croker's speech also reveals the extent to which, in its expansionary moment, the Board of Longitude almost became the Admiralty's scientific wing.

Nominally, at least, the Royal Observatory and Hydrographic Office now came under the Board's jurisdiction. It was also responsible for founding the Royal Observatory at the Cape of Good Hope, intended to be the Greenwich of the southern hemisphere, in 1820. However, before the Cape Observatory was even complete, the Board was abolished in 1828. Explained as retrenchment, the real reason for its sudden disappearance was a storm being created in the astronomical community over errors appearing in the *Nautical Almanac* and anger over how the Board's patronage was dispensed. It was initially replaced by a committee of three named scientific advisers to the Admiralty but much of the impetus for scientific work ultimately passed to the Hydrographic Office. By the 1830s, this, the observatories and the Royal Naval College were part of a defined Scientific Branch.

Both on land, therefore, and increasingly at sea, science came to play an important, though still minor, role for the Navy, through exploration and survey. Conversely, the Admiralty played a major role in the development of science in the nineteenth century. Mathematically and scientifically trained officers were prominent members of the newly founded scientific societies of the era: the Royal Astronomical Society, the Geological Society, the Zoological Society, the Royal Geographical Society and the British Association for the Advancement of Science. Naval voyages and transport proved to be vital resources for the collection of data and the development of a whole range of disciplines, from geology to botany, geography to meteorology, tidology to geomagnetism. They had potential practical benefit to Navy and nation, and a transformative effect on the shape of British science.

The first 'big science' push of the century came with the polar expeditions that were sent out in 1818, 1819 and the 1820s. The possibility of finding a North-West Passage – and the hope that they would open up new trade routes and/or find new tradable resources – joined a whole suite of navigational and scientific objectives, including magnetic and gravimetric observations. Consciously picking up on the illustrious legacy of Cook's voyages, the four ships on the 1818 expedition carried a large amount of technical and trial equipment, as well as men with considerable naval and scientific expertise. The *Isabella* and the *Alexander*, commanded by Captain John Ross and Lieutenant Edward Parry, respectively, headed north-west towards Baffin Bay. The *Dorothea* and the *Trent*, under Captain David Buchan and Lieutenant John Franklin, went due north of Spitsbergen, aiming for the North Pole. Quite specifically, national pride was at stake too. As John Barrow wrote:

> It would ... have been something worse than indifference, if, in a reign which stands proudly pre-eminent for the spirit in which voyages of discovery have been conducted, England had quietly looked on, and suffered another nation to accomplish almost the only interesting discovery that remains to be made in geography, and on to which her old navigators were the first to open the way.[8]

Barrow boasted that 'A number of new and valuable instruments were prepared for making observations in all the departments of science, and for conducting philosophical experiments and investigations'. Each expedition, he said, was supplied with a clock and a transit instrument, plus:

> a dipping needle on a new construction ... – an azimuth compass improved by Captain Kater – a repeating circle for taking terrestrial angles – an instrument for ascertaining the altitude of celestial bodies when the horizon is obscured by fogs ...– a dip-micrometer and dip-sector, invented by Doctor Wollaston ... – a macrometer ... for measuring directly the distance of inaccessible objects ... – three chronometers to each ship – a hydrometer ... – thermometers of various kinds – a barometer of Sir Henry Englefield's construction ...[9]

and more, not to mention everything that might possibly be required for collecting specimens and making drawings.

This expedition succeeded in opening a whaling route and in bringing back collections and information about the people,

Fig. 6 – 'Landing the Treasures, or Results of the Polar Expedition!!!', by George Cruikshank, 1819. This caricature shows the return of the crew of an Arctic voyage, parading their spoils, including newly-found flora and fauna as well as Jack Frost and the North Pole itself. They are being brought to the British Museum at the left of the picture, the curators of which are jumping with excitement

animals and botany of the Arctic, even if not Jack Frost and the North Pole, as suggested in Cruikshank's caricature of 'Landing the Treasures, or Results of the Polar Expedition!!!' (Fig. 6). They also inaugurated an age of polar exploration and science in which the names of Edward Parry, John and James Clark Ross (who were uncle and nephew), and John Franklin loomed large (Fig. 7). The most successful of the subsequent voyages was led by Parry, who passed through Lancaster Sound, Barrow Strait and Melville Island in 1819–20. Parry and his officers and men on the *Hecla* and the *Griper* were the only ones to receive one of the Board of Longitude's Arctic rewards. On 4 September 1819, 'At a quarter-past nine P.M., we had the satisfaction of crossing the meridian of 110° west from Greenwich, in the latitude of 74° 44' 20"; by which His Majesty's ships, under my orders, became entitled to the sum of five thousand pounds'.[10]

However, as well as being significant for reaching extremes of geographical location and enduring an icy winter, these expeditions were important for their emphasis on scientific activity and rhetoric. The combination of heroic endeavour and scientific dedication is perfectly illustrated by a portrait

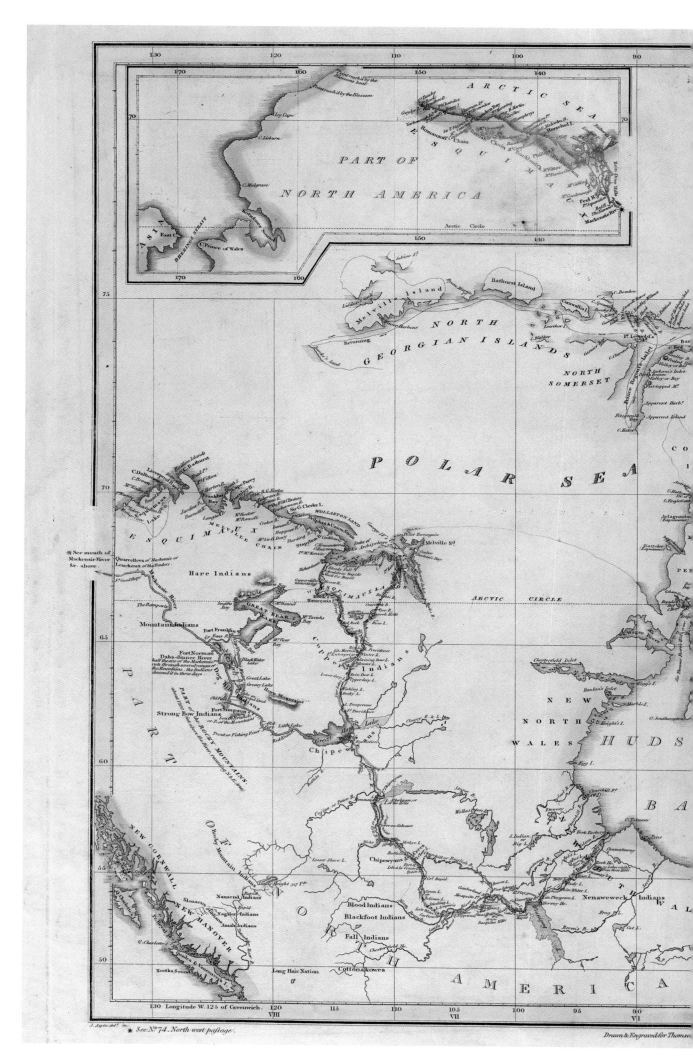

Fig. 7 – 'Discoveries of Capts Ross, Parry & Franklin in the Arctic regions from the year 1818 to 1827', from John Thomson's *New General Atlas* (London, c.1830)

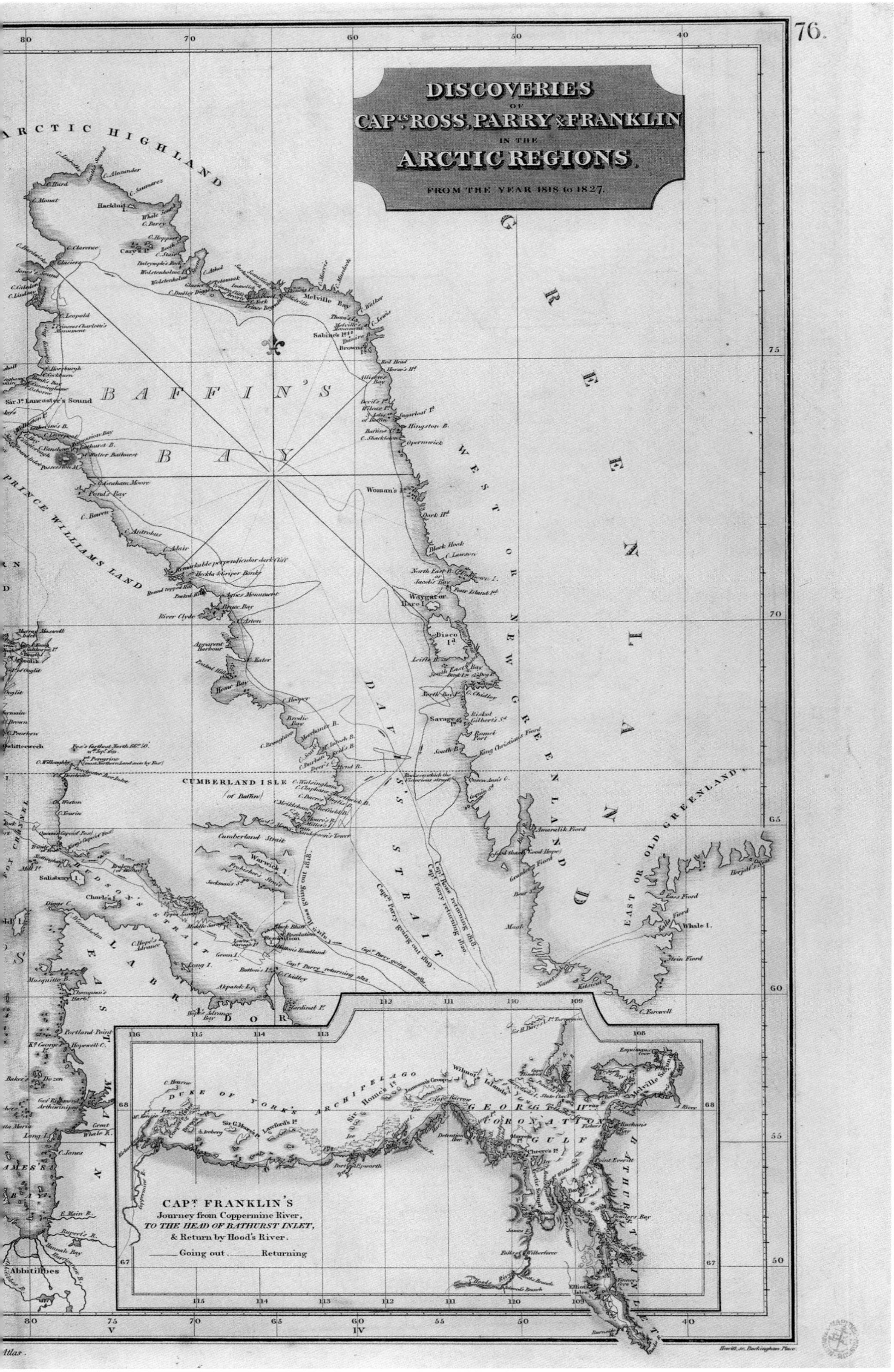
76.
DISCOVERIES
OF
CAPTS. ROSS, PARRY & FRANKLIN
IN THE
ARCTIC REGIONS.
FROM THE YEAR 1818 to 1827.
ARCTIC HIGHLAND
BAFFIN'S BAY
WEST OR NEW GREENLAND
GREENLAND
EAST OR OLD GREENLAND
PRINCE WILLIAMS LAND
CUMBERLAND ISLE (of Baffin)
DAVIS'S STRAIT
Sir Js. Lancaster's Sound
Melville Bay
Disco Id.
Cumberland Strait
Cape Farewell
CAPT. FRANKLIN'S
Journey from Coppermine River,
TO THE HEAD OF BATHURST INLET,
& Return by Hood's River.
Going out. Returning
DUKE OF YORK'S ARCHIPELAGO
GEORGE IVth CORONATION GULF
BATHURST INLET
Atlas.

of Commander James Clark Ross (Fig. 8), painted after his return from the Arctic expedition of 1829–33, in which he had succeeded in locating the north magnetic pole. It was exhibited in 1834 at the Society of British Artists as a portrait of the 'Discoverer of the North Pole'. He is painted with Romantic looks – he was, according to John Franklin's wife, 'the handsomest man in the Navy' – a sword and gleaming uniform against a black sky and icy background. He wears a bearskin, suggesting man's taming of nature and triumph over the Arctic. Just as prominent are the Pole Star and a dip circle, used to measure magnetic inclination.

Charting Africa

While polar expeditions began to fill in some of the blank spaces left on maps, there were many other areas for which charts existed but were considered inadequate. Unsurprisingly, survey voyages often focused on areas that were particularly dangerous or complex to navigate, as well as those which were strategically useful in military or trade contexts.

Somewhat different circumstances led to the charting of African coasts. In 1807, the British Parliament passed an Act designed to end the nation's participation in the slave trade, and enforcing the ban became one of the duties of the Royal Navy. The Act stated that offending ships would be seized and fines enforced. It also offered prize money to those who successfully captured such ships. Individual incentives did not convince the Navy to commit much manpower to the policing of illegal trade and one of the problems facing such missions was a lack of good information about some of the key coastlines. The eastern coast of Africa was, though, to become of increasing concern once Britain took control of Mozambique in 1810 and began to strengthen its control of Indian Ocean traffic.

The task of charting this coast fell to Captain William Fitzwilliam Owen (1774–1857), who already had considerable surveying experience and had learned some of his theory from Matthew Flinders, with whom he had been held prisoner on Mauritius between 1808 and 1810. After successful surveys of Indian Ocean islands and of the North American Great Lakes, Owen was appointed in 1821 to produce charts of Africa's east coast between the boundary of Cape Colony in present-day South Africa and Cape Guardafui in present-day Somalia. Between 1822 and 1825, Owen's squadron, formed of the sloop *Leven* and the brig *Barracouta*, surveyed around 20,000 miles of coast, which was later represented in almost 300 charts. It had been an arduous business, particularly because of the risk of fever: some seventy per cent of his officers died from disease.

Fig. 8 – Commander James Clark Ross, by John R. Wildman, 1834

Fig. 9 – Marine chronometer, by George Margetts, c.1790, used on the African surveys

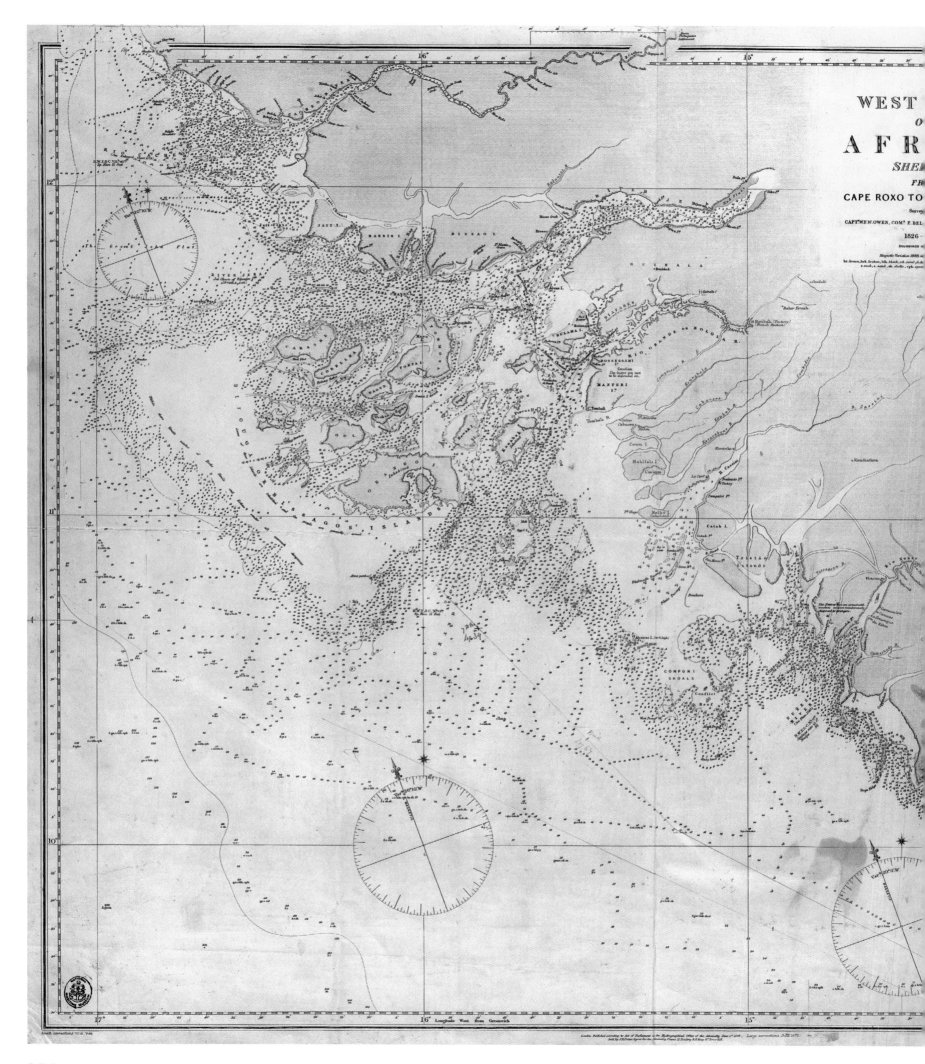

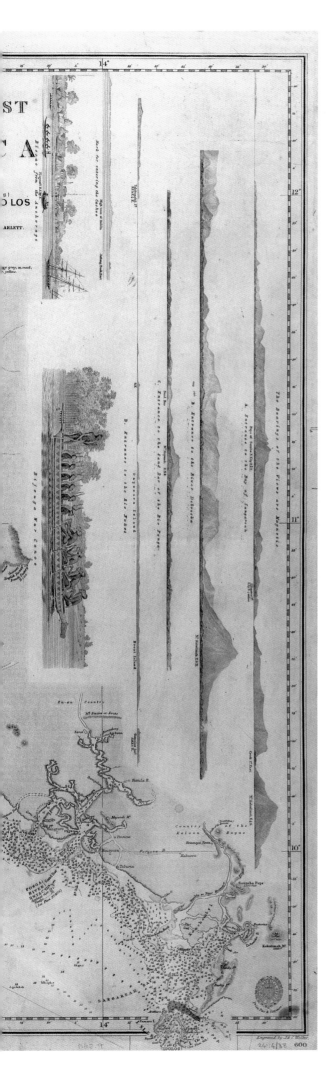

Fig. 10 – 'West Coast of Africa, Sheet VII, ... surveyed by Captain W. F. W. Owen, Commander E. Belcher, and Lieutenant W. Arlett', 1836

Fig. 11 – Advertisement for Price's candles, *c.*1850

In 1825, on their return from the East Africa survey, Owen's ships also carried out extensive surveys of the west coast of Africa (Fig. 10). They used a mixed bag of techniques on both surveys, with astronomical instruments, chronometers (Fig. 9) and, frequently, rockets all being used to fix positions and correct earlier data. Owen made much of the care and precautions taken to obtain the desired accuracy. He noted, for example, that good measurement of longitude 'would depend on the uniformity of the rates of our chronometers' and so 'we made a resolution never to fire any of the great guns from the deck on which they were kept, if it could be avoided, nor indeed from either ship any at all, unless in cases where the public service absolutely demanded it'.[11]

Specialist astronomers and a botanist were on the voyage, but Owen complained that 'Although we had a great many young officers, yet in astronomical science most of them were mere novices'. He therefore spent some time while they were stationed in Mozambique to rate the chronometers, enforcing 'a continued course of observations, both by day and night ... principally with a view to acquire the use of the different instruments'.[12]

He felt his patience and caution here were fully justified, particularly in having a sceptical approach to the trustworthiness of the timekeepers.

> Without considering the great improvements which have taken place in this instrument, and its supposed perfection, Captain Owen felt that, to place implicit confidence in it, might probably be fatal to the correctness and utility of our work: and the result proved the justice of this supposition, for, not one of our nine chronometers kept its rate without fluctuation, produced either by change of weather, climate, or position.[13]

While 'some of this variation was very trifling', it was 'in all sufficient to produce much error, unless corrected by a great deal of care and attention'.[14]

The experience gained in the use of chronometers on this voyage was shared in an 'Essay on Chronometers' by Owen's nephew, Richard. They both worried that,

Fig. 12 – The *Beagle* in the Beagle Channel, Tierra del Fuego, by Conrad Martens, *c.*1834

like many other new inventions, too much is perhaps expected from them, without paying that care and attention necessary to detect and guard against their wanderings, and those imperfections to which, as productions of human art, they must ever be liable.[15]

Setting out full instructions for their use, placement, rating and care, Richard Owen hoped to remedy the lack of this knowledge within the Navy, for, he said, the 'consequence of their falling into the hands of persons ill instructed to their use, has in many cases proved fatal'.[16]

The work in East Africa had revealed to William Owen the ongoing extent of the slave trade. Strongly religious, he saw the eradication of 'this hell-born traffic in slaves' as a Christian duty.[17] In an action undertaken on his own initiative and later disowned by the British government, Owen took the town of Mombasa, then under siege by the Sultan of Oman, and, in exchange for a promise to abolish slavery, claimed to place it under British protection. On a later voyage, he also attempted to establish a colony for free slaves on the island of Fernando Po (now known as Bioko) but was thwarted by the fever endemic there.

Charts proved helpful in other approaches to countering the slave trade, including the use of steam vessels for river patrols. They also helped the growth of alternative trades that were being encouraged by presenting them as more lucrative, as well as safer and less objectionable ethically, than the slave trade. Palm oil was a West African product promoted in this way; it was used in Britain from the 1800s, first to make soap and then, replacing tallow, to make candles after a new process was developed in the 1820s. An 1840s advertisement from the company that pioneered this product, Price's Patent Candle Company, based in Lambeth and Battersea, played out the drama for its potential customers (Fig. 11). Buying the new cleaner candles, and thus promoting the palm oil trade, was presented as a direct challenge to slavers.

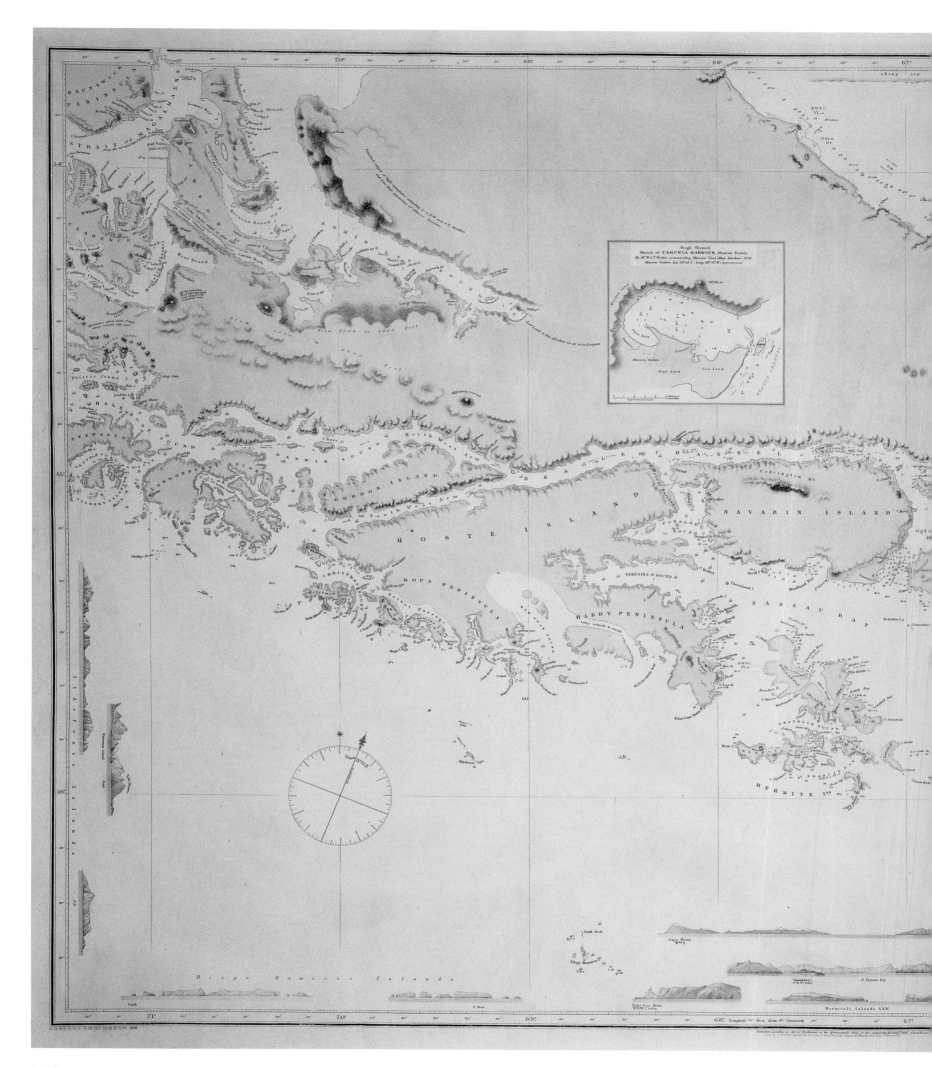

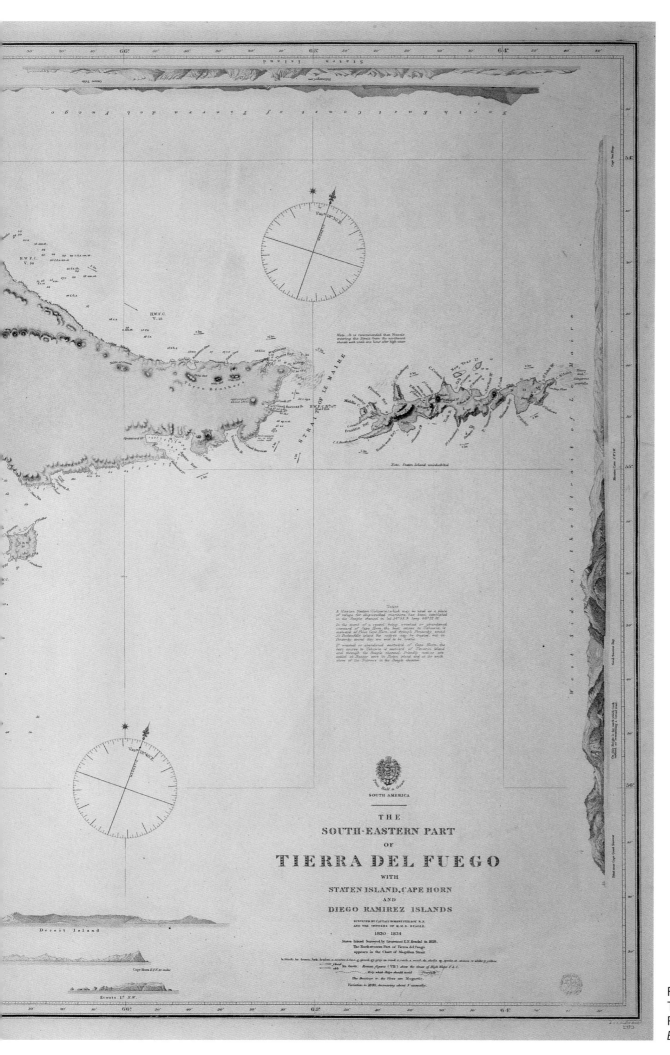

Fig. 13 – The south-eastern part of Tierra del Fuego, surveyed by Captain Robert FitzRoy and the officers of HMS *Beagle*, 1830–34

The voyages of the *Beagle*

Another area in which dedicated survey vessels were active was around Cape Horn, where the weather was dangerous and the coast rugged and complex. The most famous of all nineteenth-century survey vessels, HMS *Beagle* (Fig. 12), under Commander Pringle Stokes, headed there in 1826 with HMS *Adventure*, commanded by Captain Philip Parker King. In the difficult waters of Tierra del Fuego, Stokes committed suicide and Robert FitzRoy, the mathematically gifted naval cadet who was now a veteran of South American waters, took command of the *Beagle*.

The instructions issued to Captain King ordered him to survey the coasts of Patagonia and Tierra del Fuego (Fig. 13), and also

> to avail yourself of every opportunity of collecting and preserving Specimens of such objects of Natural History as may be new, rare, or interesting; and you are to instruct Captain Stokes, and all the other Officers, to use their best diligence in increasing the Collections in each ship: the whole of which must be understood to belong to the Public.[18]

Fig. 14 – Portrait Cove, Beagle Channel, by Conrad Martens, c.1834

Fig. 15 – Parallel rule given to John Lort Stokes by Robert FitzRoy, made by Nathaniel Worthington, c.1836. The engraved date, 1833, perhaps commemorates an important moment on their voyage, possibly Stokes's 21st birthday (detail)

The *Beagle* is best remembered today for its second voyage, on which the young Charles Darwin (1809–82) travelled as a geologist and naturalist. He and the artist Augustus Earle were taken on board by FitzRoy as private individuals, although the Navy paid for their food. Earle was later replaced by Conrad Martens, whose watercolours recorded the landscapes and people encountered (Fig. 14). While surveying took precedence, this, like other voyages, provided the opportunity for a whole range of information gathering and scientific work.

Hydrographic surveying was FitzRoy's specialism and he led the work, as well as commanding and training the officers that shared the load. Those focusing on the survey were excused the usual duties of a naval vessel. One of the most talented was the young midshipman John Lort Stokes (1812–85), who became Assistant Surveyor and was, it seems, particularly close to FitzRoy. Possibly to mark retrospectively his twenty-first birthday and, the year after, ten years in the Navy, FitzRoy gave Stokes two high-quality parallel rules (Fig. 15) and a sextant (see Fig. 17), inscribed 'Presented to J. Lort Stokes by Captn. R. FitzRoy'. Stokes received his technical education at sea, although he was able to introduce FitzRoy to a novel form of notation, which was said to be 'very convenient and assists the memory more than any other'.[19]

Stokes was to serve on all three of the *Beagle's* survey voyages, in 1826–30, 1831–36 and 1837–43, taking in the dangerous though strategically important areas of Cape Horn and Western Australia (Fig. 16). As he later wrote, over those eighteen years 'my old friend', the *Beagle*, had 'extensively contributed to our geographical knowledge'. Stokes explicitly compared this record with that of a previous *Beagle* (1806) 'that reaped golden opinions from her success in prize-making' during the war.[20] It was perhaps a plea that those serving on such scientific voyages should receive similar recognition to those involved in successful action. Stokes became lieutenant in 1837 but, even though he took command of the *Beagle* in 1841 during its third voyage, when Commander John Clements Wickham was invalided home, he was not promoted to captain until 1846.

On these voyages, Stokes naturally became familiar with a whole array of instruments, several of which would have been unknown to officers a century before (Fig. 17). On her second voyage, the *Beagle* carried no less than twenty-two chronometers, underlining the huge difference between specialist voyages for

FINDING A SHIP'S POSITION AT SEA

To find latitude

1. Measure the Sun's meridian altitude, its maximum height above the horizon around local noon. (1), (2)

2. Use the tables to calculate the Sun's true altitude, from which the latitude is derived. (3)

To find longitude

1. **Find local time**
   Measure the altitude of the Sun or a star, timed with a watch. (1), (2), (4)

   Use the tables in the *Nautical Almanac* to calculate local time and convert it to local mean time. (5), (6)

2. **Find Greenwich time**
   a) By lunar distance (7), (8), (9)
   As near simultaneously as possible, measure:
   - the angular distance of the Moon from the Sun or a star
   - the altitude of the Moon above the horizon
   - the altitude of the Sun or star above the horizon

   All observations are timed with a watch. (4)

   Use the *Nautical Almanac* to convert the measured lunar distance to the true lunar distance, calculating its value as if viewed from the Earth's centre. (10), (11)

   Look up the true lunar distance in the *Nautical Almanac* to find the equivalent time at Greenwich. (12)

   b) By chronometer (13)
   Take the time from the timekeeper and correct for its rate, the amount it gains or loses each day.

3. **Calculate the ship's longitude**
   Work out the difference between local (ship) time and Greenwich time. (11)

   Convert the time difference into an angle, one hour of time difference being equivalent to 15° of longitude.

   If Greenwich time is greater than local time, the longitude is west of Greenwich. If Greenwich time is less than local time, the longitude is east of Greenwich.

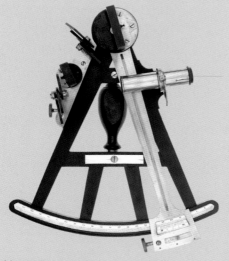

(1) Octant, 1860

(2) Sketch from the log of the *Owen Glendower*, by John Lawrence King, 1846–47

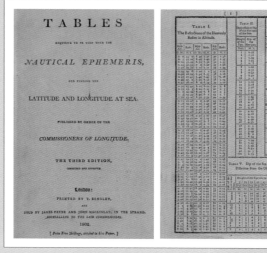

(3) *Tables Requisite*, 1802

(4) Pocket chronometer, 1778

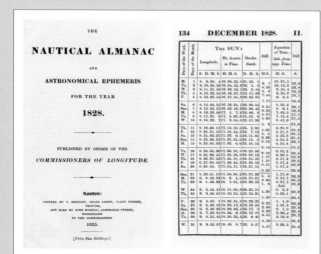

(5) *Nautical Almanac*, 1828

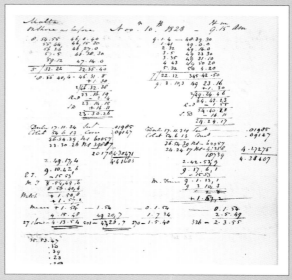

(6) Calculations, 1828

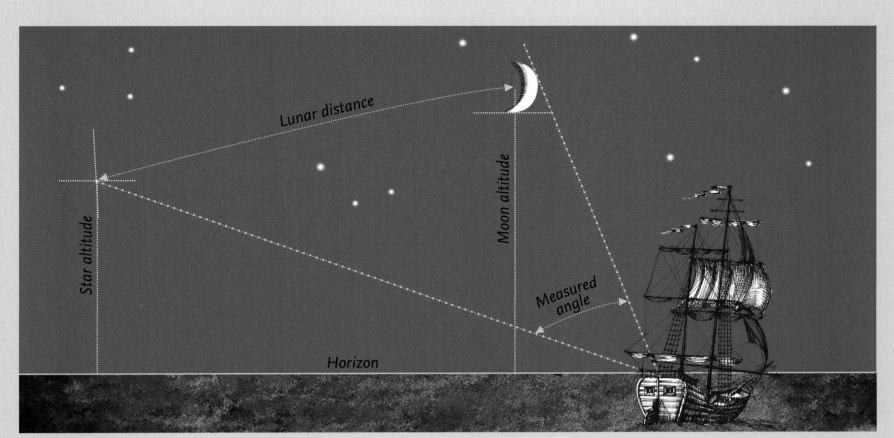

(7) Using lunar distance

(8) Sketch from the log of the *Owen Glendower*, by John Lawrence King, 1846–47

(9) Sextant, c.1840

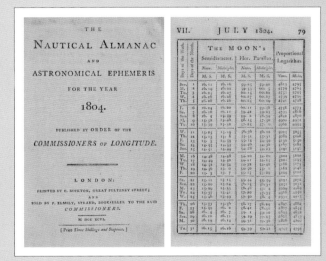

(10) *Nautical Almanac*, 1804

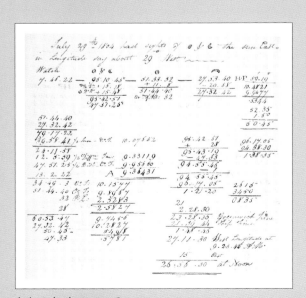

(11) Calculations, 1804

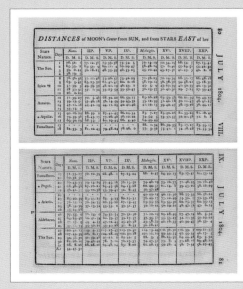

(12) *Nautical Almanac*, 1804

(13) Marine chronometer

213

hydrographic surveying and most other naval activities. These did not all belong to the Navy, nor were they all simply for ascertaining the ship's longitude – some were intended to be carried in small boats on survey activity and some were taken to be experimented on or tested themselves (Fig. 18).

Nevertheless, with so many to hand, the establishment of longitudes by chronometer was now seen as reliable enough for fixing positions on land. When discussing the difference in longitude between Rio de Janeiro and Montevideo with Francis Beaufort (1774–1857), Hydrographer to the Navy, FitzRoy noted with pleasure that not only did his results agree closely with those of Captain King on the earlier voyage, but also that observations and calculations made by the Astronomers Royal at the Cape Observatory 'confirm the *Beagle's* Chronometers'.[21] On his Australian survey, FitzRoy was instructed to check his chronometers at Parramatta Observatory, near Sydney, and then use them to determine the longitudes of various Pacific islands. This way, Beaufort said, 'all those intervening islands will become standard points to which future casual voyagers will be able to refer their discoveries or correct their chronometers'. FitzRoy felt confident that he had created 'a connected chain of meridian distances around the globe, the first that has ever been completed, or even attempted, by means of chronometers alone'.[22]

The timekeepers were, as always, used in tandem with other instruments and methods. The importance of the *Nautical Almanac* was clear when FitzRoy thanked Beaufort for sending him copies of the next year's volume: 'I was in dire alarm, thinking they would not arrive in time'. He added, too, that 'My new Sextant (with an extra glass) answers extremely well, and is a general favourite. I can take *back* observation sights for *time* when the Sun is only 22° high, and agree exactly with sights taken at the *opposite* horizon by other observers'.[23]

Despite all the brand-new technologies at their disposal, the old enemies of weather and treacherous coasts remained a danger, especially when rounding the Horn. FitzRoy wrote, 'I assure you that this last cruise has rendered me an implicit believer in all that is said in Lord Anson's Voyage, and previously I

Fig. 16 – The Victoria River on Australia's north-western coast, surveyed by John Lort Stokes in 1839; published 1845

Fig. 17 – Theodolite by J. Dollond & Son, *c.*1840, and sextant by Worthington & Allen, *c.*1831. Both belonged to John Lort Stokes, the latter being presented by Robert FitzRoy

considered that account exaggerated'.²⁴ Such shores and weather made surveying particularly challenging and, before any day's observations could begin, they had to find safe harbour. Only then could 'observations for Latitude, time, and true bearings, on the Tides and Magnetisation' be attempted, and plans drawn up. These plans were based on triangulation, which involved measuring the angles made between significant features of the landscape and a base line connecting two temporary observatories or hill-top stations.²⁵

'Many leagues of exposed and difficult coast', FitzRoy wrote, 'were looked down upon in this manner, and at the least their exact bearings from one fixed spot ascertained.' Ideally, the length of the base line used in the triangulation would be determined by one of four methods. In descending order of value, they were, 'deduced from good Astronomical or Chronometrical observations'; 'deduced from angular measurements of small spaces exactly known'; 'obtained by actual measurement with a chain, with rods or with a line'; or, 'obtained by sound', that is, rockets.²⁶

FitzRoy emphasized accuracy but also made clear that hair-splitting levels of precision, would not make a difference to the detail available on a chart, which could be sacrificed for efficiency. He concluded:

> By multiplying bases, which with such easy methods is soon effected; and by a frequent use of the sextant, artificial Horizon and Chronometer, material errors may be kept out of the work of a practical surveyor. With a Sextant, Horizon,

Fig. 18 – Marine chronometer no. 294, by John Roger Arnold, c.1807, used on the *Beagle* and later converted into a travel clock

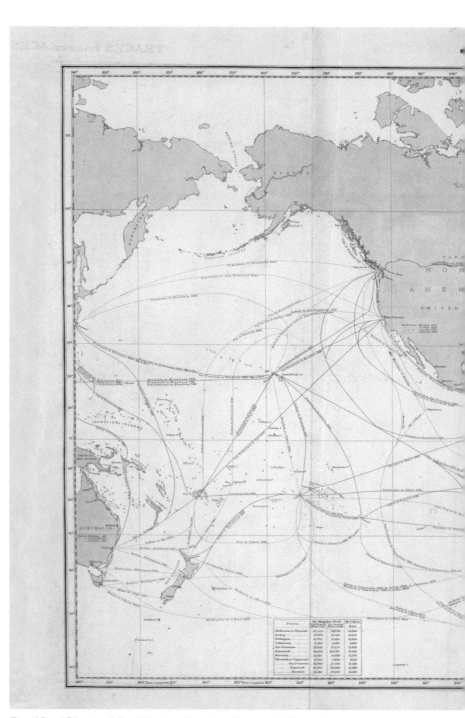

Fig. 19 – 'Chart of the World, showing Tracks followed by Sailing and Low Powered Steam Vessels', 1888

and Chronometer (in a sheltered spot), a micrometer ... a Theodolite, and intelligent Assistants, much work may be done in a short time.[27]

The *Beagle's* and other surveys of the nineteenth century were extraordinary, dangerous and painstaking endeavours. They achieved new levels of precision in an ever-increasing series of charts and added to the stock of data and specimens in a whole range of scientific fields. They helped underpin the nation's trade and enabled further exploitation of the resources of a growing empire. Even where the current value of particular territories was as yet unclear, there was confidence that increased knowledge would render them profitable. As Stokes wrote of Australia's unpromising northern territory,

no one who reflects on the power of trade to knit together even more distant points of the earth, will think it visionary to suppose that Victoria [Settlement] must one day – insignificant as may be the value of the districts in its immediate neighbourhood – be the centre of a vast system of commerce ...[28]

Stokes was wrong about the potential of the recently founded Port Essington (Victoria Settlement), which he imagined as 'the emporium ... where will take place the exchange of the products of the Indian Archipelago for those of the vast plains of Australia', for it was abandoned in 1849. However, his comments speak for the sentiments that lay behind much of the scientific effort of the British Admiralty and the Royal Navy.[29]

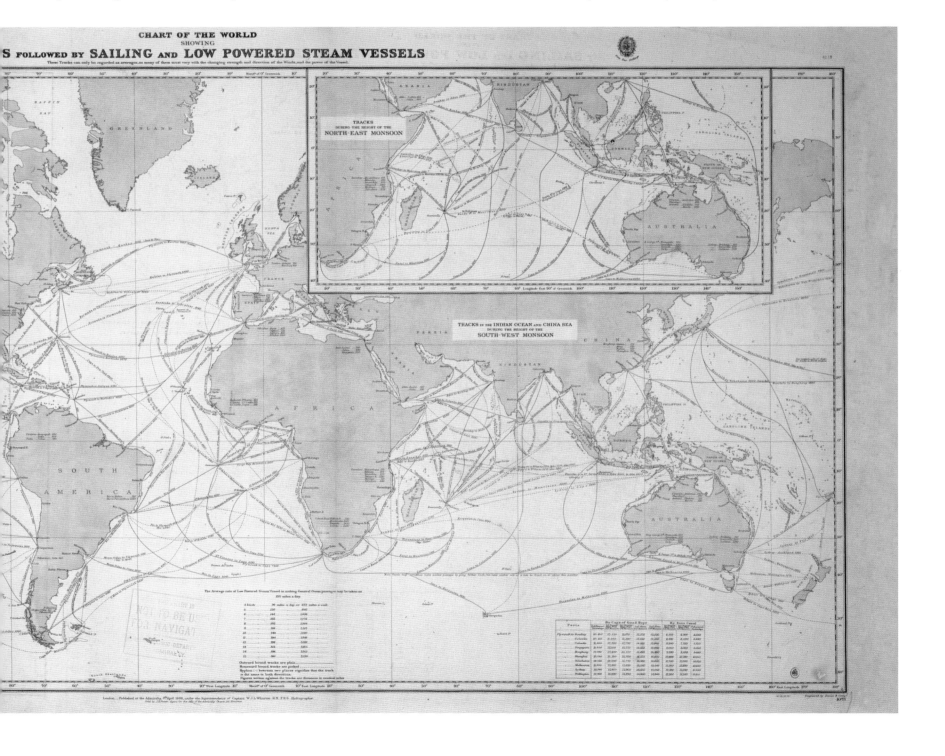

Fig. 20 – Visualization of international shipping routes, based on the Climatological Database for the World's Oceans 1800–1850 (CLIWOC)

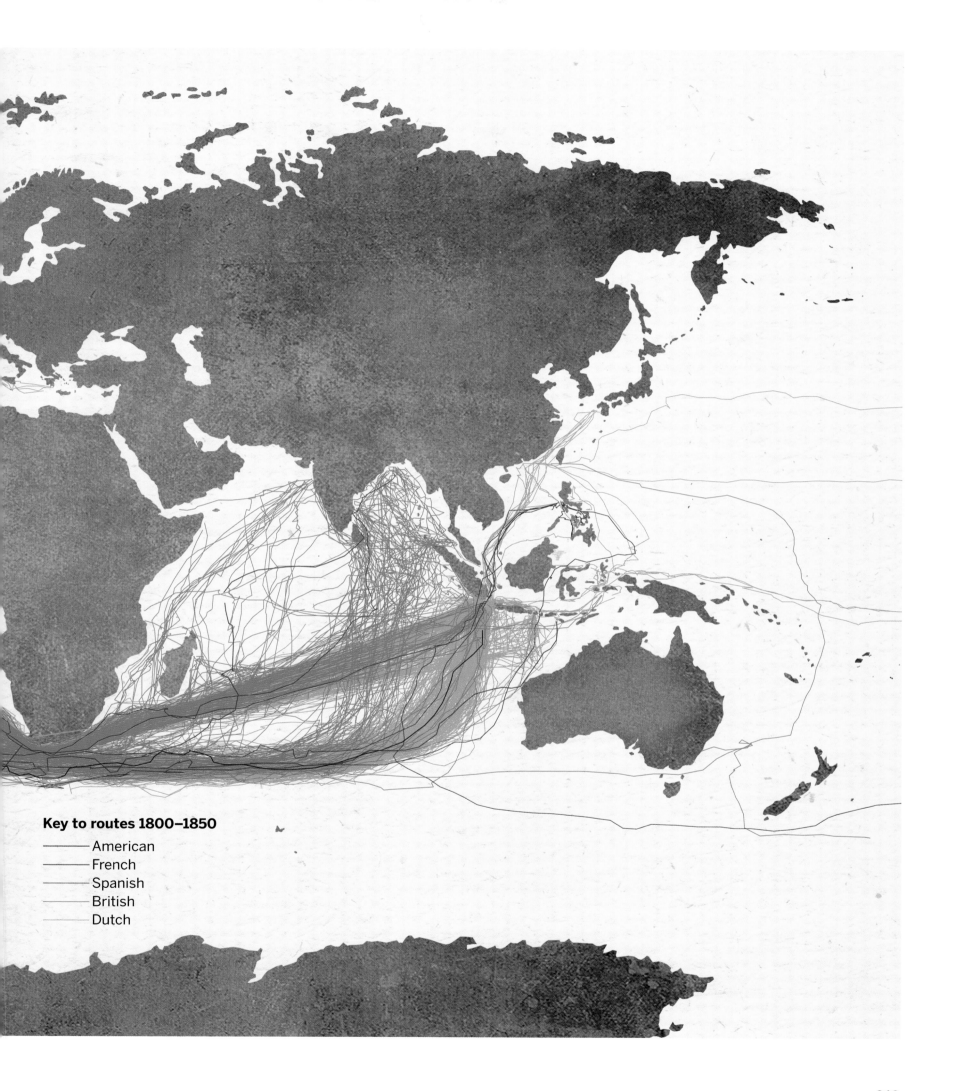

Key to routes 1800–1850

— American
— French
— Spanish
— British
— Dutch

In 1836, the Hydrographer of the Navy, Francis Beaufort, wrote of how,

Every day, intelligence arrives of the discovery of new rocks and shoals, or of more correct limits being assigned to those already known. Improved means of observation are constantly yielding more accurate positions in lat.^{de} and long.^{de}. The time and height of Harbour Tides, the set and velocity of sea currents, the altitudes of conspicuous headlands, descriptions of foreign ports, with the modes of obtaining water, wood and refreshments are scattered in expensive books. Lighthouses are springing up in various parts of the world, and at home continual changes are making in the places of buoys and beacons. And new charts and sailing directions are frequently published as well as gradual improvements in the practice of navigation ...[30]

This was a world that was, in many ways, shrinking and increasingly coming under the surveillance and control of metropolitan elites. Scientific methods and institutions were at the heart of this. Observatories were aiding the use of chronometers and providing 'a local centre of reference' for navigation, cartography and territorial administration. Each was, said the astronomer John Herschel, in his presidential address

to the British Association for the Advancement of Science, 'a nucleus for the formation around it of a school of exact practice', almost synonymous with civilization itself.[31] Nothing before the late twentieth century defined a position or the local time more exactly than an observatory.

This demand for precision, as well as the worldwide use of charts and tables based on the observations made at the Royal Observatory, Greenwich, ultimately led to the Greenwich meridian serving as the International Prime Meridian. It had slowly become the standard reference point by which to determine longitude, a fact recognised by its proposal as the world's common starting place for measuring time and longitude by the delegates at the International Meridian Conference held in Washington D.C. in 1884. Over the course of the nineteenth century, chronometers, backed by astronomical methods and served by observatories, likewise became the standard means of keeping track of longitudinal position at sea and a key tool for surveys. The legacy of the search for longitude methods, and their application, together with innovations such as steam power and the electric telegraph, remains in the mapping of the world and in the patterns that were formed, connecting trade, empires and people (Figs 19 and 20).

Epilogue

O'er the glad waters of the dark blue sea,
Our thoughts as boundless, and our souls as free,
Far as the breeze can bear, the billows foam,
Survey our empire, and behold our home!

Byron, The Corsair, Canto I[1]

This story of the search for a way to find longitude at sea has taken us through several centuries and around the world. Monarchs, mathematicians, artisans, businessmen, politicians and mariners have all played their roles in defining the problem, looking for solutions and slowly making them part of life at sea. Many of the risks of sea travel remained but keeping track of, or finding, a ship's longitude became possible for any well-supplied vessel with a trained crew. Crucially, this position could be tracked on reliable charts, which themselves had been made possible by the infrastructure that had supported the longitude solutions – a thriving instrument trade and government investment in astronomical observatories and specialist skills.

These milieus revealed the abilities and dedicated effort of some remarkable individuals. Many well-known figures in the history of science – Galileo, Huygens, Newton, Halley – helped explore possible avenues of research, but others, from all walks of life, had to enter the story to turn theory into practice. Some names have, of course, loomed larger than others in the preceding pages. John Harrison's remarkable abilities, for example, led to a career unique in eighteenth-century Britain: by receiving government support he was able to focus on just one problem over three decades. However, the time was ripe for such inventions. Other clock- and watchmakers, in Britain and in France, were also developing the necessary skills, materials and processes. It was their combined efforts that developed the marine chronometer.

Astronomy, always the closest ally of horology, was an even more international affair. Newton's theories were improved by French mathematicians and applied by the Hanoverian Tobias Mayer to tame the complex theory of the Moon's motions sufficiently to meet the purposes of navigators. While the French showed how these numbers could be processed to aid their use at sea, it was the head of the Royal Observatory, backed by a government board, who established a system that made the production of such tables a matter of ongoing routine. Nevil Maskelyne's *Nautical Almanac* was widely copied and put to use around the world, supporting all methods of navigation. Maskelyne himself did much to ensure the development and use of marine timekeepers, and his successors as Astronomer Royal added supervision of the Royal Navy's chronometers and time signals to their daily business.

By the nineteenth century, the tools existed to allow naval officers and other travellers to chart their courses accurately and to carry out detailed surveys. The world slowly became better mapped and defined. This was for good and ill. While improved navigation and charts helped reduce the risk of sea voyages, they also supported military and imperial endeavour, and the exploitation of many people and resources for the profit of a few.

It was not until the twentieth-century advent of, first, wireless radio signals and, later, positioning systems including satellite navigation, that these nineteenth-century methods became obsolete. On 1 November 1968, it was decreed that ships of the Royal Navy should cease to carry marine chronometers, although sextants, almanacs and, of course, dead reckoning remained as a back-up to electronic devices. Given the potential vulnerability of a military-backed system like the Global Positioning System (GPS), it is worth remembering that, should a ship in open water lose communication and its longitude, it is only by astronomy that the latter can be re-established.

It is more than likely that the new longitude techniques would have been developed without the passing of the Longitude Act of 1714, for other rewards or the possibility of developing a profitable business were reason enough to invest in potential solutions. The Act did, however, help cement a lasting alliance between the scientific academy, comprising Oxford and Cambridge universities and the Royal Society, and government. Today, government investment in science and technology, while often appearing under threat, has grown beyond anything that could have been imagined at the start of the nineteenth century. This has not stopped a revival of interest in the idea of retrospectively rewarding those who solve defined problems. Challenge prizes have proliferated since the creation of the XPRIZE Foundation, which sets up competitions and offers large prizes to the winners. The first X Prize was offered to the first non-governmental organization that could launch a manned spacecraft twice within two weeks. It was won in 2004, with the $10 million prize having inspired an estimated total investment of $100 million by the twenty-six competing teams.

This year, 2014, will see the launch of a new prize that is partly funded by the British government. At the time of writing it is being promoted as a new Longitude Prize, although the issues it will attempt to tackle are yet to be decided. Unlike the original backers of the 1714 Longitude Act, the organizers also hope to gauge public opinion on what questions might be answered, although ultimately the range of possible problems and potential solutions will be defined by scientists and politicians. Our story here has shown that, whatever the topic, it will be necessary to outline the terms of the prize with care and to develop the infrastructure required to put new ideas into regular practice.

The effort to map and understand our world – to measure it, its resources and populations – was begun in earnest in the nineteenth century and continues today. It is an inherently political process but, as well as revealing opportunities to gain power and profit, it should also help us to identify and perhaps solve some of the problems that we face on a planet both seemingly smaller and undoubtedly more stressed than it was in 1714.

(following page) – 'The *Great Western* riding a tidal wave, 11 December 1844', by Joseph Walter, 1845

CHAPTER 1

Fig. 1

Carte universelle du commerce, by Pierre Du-Val, Paris, 1686 (NMM G201:1/32)

Fig. 2

A busy Dutch East Indies factory port, possibly Surat, by Ludolf Backhuysen, 1670 (NMM BHC1933)

Fig. 3

'A description of the old town & the port of realejo' (El Viejo, Nicaragua), from 'A Waggoner of the South Sea', by William Hack, 1685 (NMM P/33)

Fig. 4

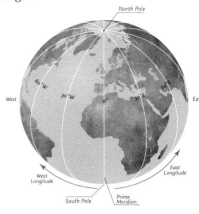

Longitude lines

Fig. 5

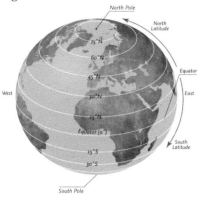

Latitude lines

Fig. 6

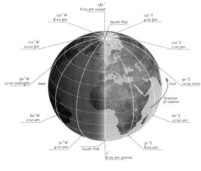

Time and longitude

Fig. 7

A mariner's compass made by Jonathan Eade in London, *c*.1750 (NMM NAV0378)

Fig. 8

Log of the *Orford*, by Lieutenant Lochard, October 1707 (NMM ADM/L/O/22)

Fig. 9

A backstaff, by Will Garner, London, 1734 (NMM NAV0041)

Fig. 10

A seaman with a lead and line, from *The Great and Newly Enlarged Sea Atlas or Waterworld*, by Johannes van Keulen (Amsterdam, 1682) (NMM PBD8037)

Fig. 11

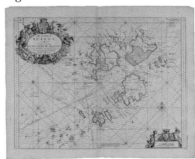

'The Islands of Scilly', from *Great Britain's Coasting Pilot*, by Greenvile Collins (London, 1693) (NMM PBD8205)

Fig. 12

The Indian Ocean, from *The Great and Newly Enlarged Sea Atlas* (Amsterdam, 1682) (NMM PBD8037)

Fig. 13

Practical Navigation, by John Seller (London, 1672) (NMM PBC1327)

Fig. 14

Two English ships wrecked in a storm on a rocky coast, by Willem van de Velde the Younger, *c*.1700 (NMM BHC0907)

Fig. 15

Sir Cloudisly Shovel in the Association with the Eagle, Rumney and the Firebrand, Lost on the Rocks of Scilly, October 22, 1707 (NMM PAH0710)

1. William Bourne, *A Regiment for the Sea* (London, 1574), p. 42v.
2. *Old Bailey Proceedings Online*, December 1692, trial of John Glendon (t16921207-7) <www.oldbaileyonline.org> [accessed 4 June 2013].
3. E. Chappell (ed.), *The Tangier Papers of Samuel Pepys* (London, 1935), pp. 127–28.
4. John Flamsteed to Samuel Pepys, 21 April 1697, BL Add MS 30221, p. 187, quoted in W. E. May, *A History of Marine Navigation* (Henley, 1973), p. 29.
5. John Dryden, *Annus Mirabilis* (London, 1667), p. 42 verse 163.
6. Chappell, *Tangier Papers*, p. 130.
7. Quoted in John Gascoigne, *Captain Cook: Voyager Between Worlds* (London, 2007), pp. 53–54.
8. Benjamin Franklin, 'Journal of Occurrences in My Voyage to Philadelphia', 1726, quoted in Jonathan Raban (ed.), *The Oxford Book of the Sea* (Oxford, 1992), p. 105.
9. William Dampier, *A New Voyage Round the World*, 5th edn (London, 1703), p. 100.
10. Quoted in S. E. Morison, *Admiral of the Ocean Sea: A Life of Christopher Columbus* (Boston, 1992), p. 196.
11. William Funnell, *A Voyage Round the World. Containing an Account of Captain Dampier's Expedition into the South-Seas* (London, 1707), p. 14.
12. William Dampier, *Capt. Dampier's Vindication of his Voyage to the South-Seas* (London, 1707), p. 3.
13. Magdalene College, Cambridge, Pepys MS 2612, p. 104, quoted in N. Plumley, 'The Royal Mathematical School within Christ's Hospital: The Early Years – Its Aims and Achievements', *Vistas in Astronomy*, 20 (1976), 51–59 (p. 52).
14. Ibid., p. 56.
15. Chappell, *Tangier Papers*, p. 129.

CHAPTER 2

Fig. 1

'An Act for Providing a Publick Reward for such Person or Persons as shall Discover the Longitude at Sea' (the Longitude Act, 1714) (Parliamentary Archives HL/PO/PU/1/1713/13An35)

Fig. 2

Isaac Newton, by Charles Jervas, 1717 (The Royal Society P/0095)

Fig. 3

William Whiston, by an unknown artist, c.1690 (Clare College, Cambridge)

Fig. 4

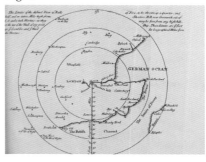

William Whiston's *The Longitude Discovered* (London, 1738) (NMM PBA2144)

Fig. 5

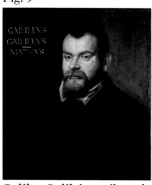

A terrella (or 'little earth'), c.1600 (NMM ACO1368)

Fig. 6

An amplitude compass, made by Ferreira, Lisbon, 1780 (NMM NAV0462)

Fig. 7

Edmond Halley, by Thomas Murray, c.1690 (The Royal Society P/0059)

Fig. 8

Edmond Halley's world sea chart on two sheets, showing lines of equal magnetic variation, 1702 (NMM G201:1/1A-B)

Fig. 9

Galileo Galilei, attributed to Francesco Apollodoro, c.1602–07 (NMM BHC2699)

Fig. 10

Galileo's journal of the observations of Jupiter and its satellites, 1610 (Biblioteca Nazionale Centrale, Florence MS Gal. 48, f. 30r)

Fig. 11

Paris Observatory, 1729 (NMM PAJ3502)

Fig. 12

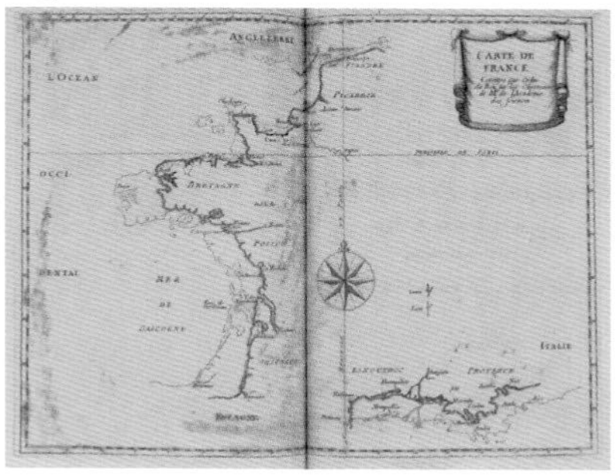

Map of France from *Recueil d'Observations* (Paris, 1693) (NMM PBG0584)

Fig. 13

Title page of *Introductio Geographica*, Peter Apian (Ingolstadt, 1533) (NMM PBN1982)

Fig. 14

Decorated ivory cross-staff, by Thomas Tuttell, c.1700 (NMM NAV0505)

Fig. 15

Royal Observatory from Crooms Hill, British School, c.1696 (NMM BHC1812)

Fig. 16

The effects of the Sun on the Moon's motion, from Isaac Newton's *Philosophiae Naturalis Principia Mathematica* (Cambridge, 1713) (NMM AAB0813)

Fig. 17

The Octagon Room at the Royal Observatory, Greenwich, by Francis Place, *c*.1676
(NMM ZBA1808)

Fig. 18

Marine timekeeper, by Severyn Oosterwijck, *c*.1662
(Private collection)

Fig. 19

Frontispiece to Thomas Sprat's *History of the Royal-Society of London* (London, 1667)
(NMM PBG0382)

Fig. 20

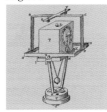

Christiaan Huygens' design for a marine timekeeper, originally drawn *c*.1685–86, from *Oeuvres Complètes de Christiaan Huygens*, 22 vols (Den Haag, 1888–1950), IX, p. 54
(Private collection)

Fig. 21

Longitude timekeeper, designed by Lothar and Conrad Zumbach de Koesfelt, made by Franciscus le Dieu, 1749
(Boerhaave Museum, Leiden V09204)

Fig. 22

Plate from Henry Sully's *Description Abrégée d'une Horlorge d'une Nouvelle Invention* (Paris, 1724)
(NMM ZBA2248.2)

[1] *An Act for providing a Publick Reward for such Person or Persons as shall discover the Longitude at Sea* (13 Anne, c. 14), 1713.

[2] 'In praise of the choice company of Philosophers and Witts who meet on Wednesdays weekly, at Gresham College', BL Add. MS 34217, ff. 30–1, quoted in F. Sherwood Taylor, 'An Early Satirical poem on the Royal Society', *Notes and Records of the Royal Society*, 5 (1947), pp. 37–46, (p. 42).

[3] Miguel de Cervantes, *A Dialogue Between Scipio and Bergansa, Two Dogs Belonging to the City of Toledo* (London, 1767), pp. 107–08.

[4] *An Act for providing a Publick Reward for such Person or Persons as shall discover the Longitude at Sea* (13 Anne, c. 14), 1713.

[5] The National Archives Currency Converter <http://www.nationalarchives.gov.uk/currency/> [accessed 29 November 2013].

[6] Isaac Newton, 'Report to the Lords of the Admiralty on the Different Projects for Determining the Longitude at Sea', Newton Papers, CUL Add.3972, p. 32r (CDL, <http://cudl.lib.cam.ac.uk/view/MS-ADD-03972/63> [accessed 3 December 2013]). The wording is almost the same as that recorded in the House of Commons Journal Book.

[7] William Whiston, *Memoirs of the Life and Writings of Mr. William Whiston* (London, 1749), p. 294.

[8] *Journals of the House of Commons*, 17 (London, 1803), 25 May 1714, pp. 641–42.

[9] Isaac Newton, 'Report to the Lords of the Admiralty'.

[10] Benedetto Castelli to Galileo, 11 September 1617, quoted in Albert Van Helden, 'Longitude and the Satellites of Jupiter', in William J. H. Andrewes (ed.), *The Quest for Longitude* (Cambridge, Mass., 1996), pp. 85–100, (p. 91).

[11] Clements R. Markham (ed.), *The Voyages of William Baffin* (London, 1881), p. 22.

[12] Quoted in D. J. Bryden, 'Magnetic Inclinatory Needles: Approved by the Royal Society?', *Notes and Records of the Royal Society of London*, 47 (1993), 17–31 (p. 19).

[13] Royal Warrant, quoted in Derek Howse, *Greenwich Time and the Longitude* (London, 1997), p. 42.

[14] John Conduitt, 'Notes on Newton's character', King's College, Cambridge, Keynes MS 130.07, 6v.

[15] William Whiston and Humphry Ditton, *A New Method for Discovering the Longitude both at Sea and Land* (London, 1714), pp. 16–17.

[16] Ibid.

CHAPTER 3

Fig. 1

An ABSTRACT of the Act of Parliament concerning the Discovery of the
LONGITUDE,
And other Improvements of NAVIGATION.

FINIS.

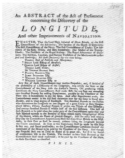

A summary of the 1714 Longitude Act
(Het Scheepvaartmuseum, Amsterdam A.4876 (546))

Fig. 2

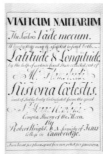

VIATICUM NAUTARUM
or
The Sailor's Vade mecum.

Latitude & Longitude

Historia Cœlestis.

Complete Theory of the Moon.
By
Robert Wright. B.A. formerly of Jesus College in Cambridge.

'Viaticum Nautarum or The Sailor's Vade Mecum', by Robert Wright, 1726
(NMM NVT/5)

Fig. 3

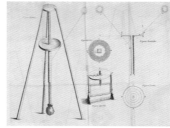

Dorotheo Alimari's observing instrument, from *The New Method Proposed by Dorotheo Alimari to Discover the Longitude*, by Sebastiano Ricci (London, *c.*1714)
(NMM PBD1145)

Fig. 4

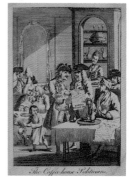

A satirical print, *The Coffee House Politicians*, *c.*1733
(British Museum 1868,0808.13254)

Fig. 5

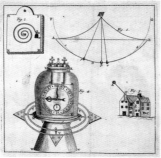

Jeremy Thacker's proposed longitude timekeeper, from *The Longitudes Examin'd* (London, 1714)
(British Library 533.f. 22)

Fig. 6

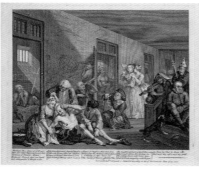

The final plate of William Hogarth's *A Rake's Progress*, 1735
(British Museum 1868,0822. 1536)

Fig. 7

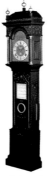

Precision long-case regulator, by John Harrison, 1726
(Private collection)

Fig. 8

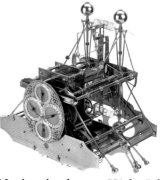

Marine timekeeper H1, by John Harrison, completed in 1735
(NMM/MoD ZAA0034)

Fig. 9

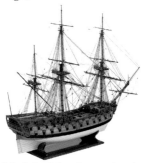

Model of the *Centurion*, by Benjamin Slade, made in 1747 for George Anson
(NMM SLR0442)

Fig. 10

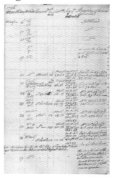

Log of the *Centurion*, by Captain John Proctor, 1736
(NMM ADM/L/C/82)

Fig. 11

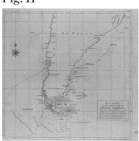

'A Chart of the Southern Part of South America with the Track of the Centurion',
by Richard Seale, 1748
(NMM G244:1/2)

Fig. 12

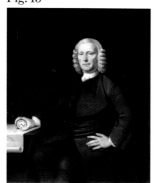

Marine timekeeper H2, by John Harrison, 1737–39
(NMM/MoD ZAA0035)

Fig. 13

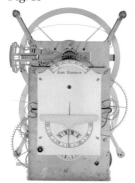

Marine timekeeper H3, by John Harrison, 1740–59
(NMM/MoD ZAA0036)

Fig. 14

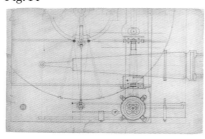

Drawing of part of H3, by John Harrison, *c.*1740
(NMM ZAA0882)

Fig. 15

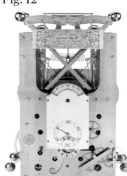

John Harrison, by Thomas King, *c.*1765–66
(Science Museum 1884-217)

Fig. 16

Construction drawings for the mechanism of H4, by John Harrison, c.1756 (Worshipful Company of Clockmakers)

Fig. 17

Marine timekeeper H4, by John and William Harrison, 1755–59 (NMM/MoD ZAA0037)

Fig. 18

John Hadley, attributed to Bartholomew Dandridge, early 1730s (NMM BHC2731)

Fig. 19

Hadley quadrant or octant, c.1744 (Het Scheepvaartmuseum, Amsterdam B.0021(02))

Fig. 20

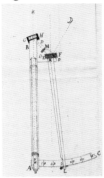

Thomas Godfrey's proposal for a double-reflection instrument (The Royal Society EL/L6/59)

Fig. 21

Backstaff with artificial horizon, designed by John Elton and made by Jonathan Sisson, c.1732 (NMM NAV0039)

Fig. 22

'Book of Drafts and Remarks', by Archibald Hamilton, 1763 (NMM NVP/11)

Fig. 23

Tobias Mayer, by an unknown artist, mid-eighteenth century (Private collection)

Fig. 24

'Germaniae atque in ea Locorum Principaliorum Mappa Critica', by Tobias Mayer, 1750 (British Library Maps 26905.(27.))

Fig. 25

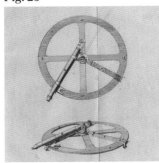

Mayer's design for a repeating circle, from *Tabulae Motuum Solis et Lunae Novae* (London, 1770) (NMM PBG0823)

Fig. 26

Marine sextant, by John Bird, c.1758 (NMM NAV1177)

Fig. 27

Nevil Maskelyne, by John Russell, c.1776 (NMM ZBA4305)

Fig. 28

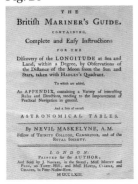

The British Mariner's Guide, by Nevil Maskelyne (London, 1763) (NMM PBD1847)

Fig. 29

A sixteenth-century design for a marine observing chair, from *Le Cosmolabe*, by Jacques Besson (Paris, 1567) (NMM PBE5169)

Fig. 30

'View of Bridgetown and part of Carlisle Bay in the Island of Barbadoes', by Edward Brenton, late eighteenth century (NMM PAF8416)

1. Quoted in Barry Slepian, 'George Faulkner's Dublin Journal and Jonathan Swift', *Library Chronicle*, 31 (1965), 97–116 (p. 100).
2. Anonymous, *A Hymn to the Chair: or, Lucubrations, Serious and Comical, on the Use of Chairs, Benches, Forms, Joint-Stools, Three-Legged Stools, and Ducking-Stools* (London, 1732), p. 19.
3. Daniel Defoe, *Essays upon Several Projects* (London, 1702), p. 4.
4. Edmund Stone, *Some Reflections on the Uncertainty of Many Astronomical and Geographical Positions* (London, 1766), pp. x–xi.
5. John Flamsteed to Abraham Sharp, 31 August 1714, in E. G. Forbes et al. (eds), *The Correspondence of John Flamsteed, The First Astronomer Royal*, 3 vols (Bristol & Philadelphia, 2002), III, p. 700.
6. John Flamsteed to Abraham Sharp, 23 October 1714, in Forbes et al., *Correspondence of John Flamsteed*, III, p. 712.
7. Isaac Newton, draft letter on finding the longitude at sea, CUL Add.3972.3–4, fol. 38r (CDL, <http://cudl.lib.cam.ac.uk/view/MS-ADD-03972/75> [accessed 3 December 2013]).
8. Jane Squire, *A Proposal to Determine Our Longitude* (London, 1743), p. 40.
9. Ibid., p. 33.
10. John Houghton, 'A Discourse of Coffee', *Philosophical Transactions*, 21 (1699), 311–17 (p. 317).
11. 'Lecture the 19th Read on Monday, 26th February 1705/6', CUL RGO 1/68/G, fols 261–62.
12. Jeremy Thacker, *The Longitudes Examin'd* (London, 1714), title page, p. iii.
13. William Ward and Caleb Smith, *The Description and Use of a New Astronomical Instrument* (London, 1735), p. 4.
14. Jonathan Swift, *Travels into Several Remote Nations of the World ... By Lemuel Gulliver*, 2 vols (London, 1726), II, pp. 133, 136.
15. George Lyttelton, *Letters from a Persian in England* (London, 1735), p. 14.
16. John Arbuthnot to Jonathan Swift, 17 July 1714, quoted in Larry Stewart, *The Rise of Public Science* (Cambridge, 1992), p. 192.
17. Charles Kirby-Miller (ed.), *Memoirs of the Extraordinary Life, Works and Discoveries of Martinus Scriblerus* (Oxford, 1988), pp. 166–67.
18. 'Ode for Musick on the Longitude', in Jonathan Swift, *A Supplement to Dr. Swift's works* (Edinburgh, 1753), pp. 341–42.
19. *The Longitude Discover'd: A Tale* (London, 1726), p. 10.
20. *The Family Memoirs of the Rev. William Stukeley M.D.*, 3 vols (Durham, 1882–87), II, p. 298.
21. Sir Charles Wager to Captain Proctor, 14 May 1736, quoted in John Harrison, *An Account of the Proceedings, in Order to the Discovery of the Longitude* (London, 1763), p. 17.
22. Captain Proctor to Sir Charles Wager, 17 May 1736, quoted in Harrison, *An Account of the Proceedings*, p. 18.
23. *London Evening Post*, 30 June 1737.
24. BoL, confirmed minutes, 4 June 1746, CUL RGO 14/5, p. 14 (CDL, <http://cudl.lib.cam.ac.uk/view/MS-RGO-00014-00005/18> [accessed 3 December 2013]).
25. William Hogarth, *The Analysis of Beauty* (London, 1753), p. 70.
26. BoL, confirmed minutes, 17 August 1762, CUL RGO 14/5, p. 38 (CDL, <http://cudl.lib.cam.ac.uk/view/MS-RGO-00014-00005/42> [accessed 3 December 2013]).
27. John Hadley, 'The Description of a New Instrument for Taking Angles', *Philosophical Transactions*, 37 (1731–32), 147–57.
28. John Hadley, 'An Account of Observations Made on Board the Chatham-Yacht, August 30th and 31st, and September 1st, 1732', *Philosophical Transactions*, 37 (1731–32), 341–56 (p. 351).
29. Lieutenant John Elliot to his father, 4 May 1756, NMM ELL/400, item 12.
30. Tobias Mayer to Carsten Niebuhr, 2 July 1761, Universitäts-Bibliothek, Christian-Albrechts Universität Kiel, Nachlass Carsten Niebuhr, Cod. MS KB 314.5, no. 8, quoted in Lawrence J. Baack, '"A Practical Skill that Was Without Equal": Carsten Niebuhr and the Navigational Astronomy of the Arabian Journey, 1761–7', *Mariner's Mirror*, 99 (2013), 138–52 (p. 143).
31. Nevil Maskelyne, *The British Mariner's Guide* (London, 1763), pp. iv–v.
32. *The London Magazine, or, Gentleman's Monthly Intelligencer*, 28 (1759), 505.
33. *Lloyd's Evening Post and British Chronicle*, 8–10 October 1760.
34. *Busy Body*, 27 October 1759.
35. Bengt Ferrner, diary entry, 12 October 1759, in Bengt Ferrner, *Resa I Europa 1758–1762*, ed. by Sten G. Lindberg (Uppsala, 1956), p. 147 (translation by Jacob Orrje).
36. William Philip Best to Johann David Michaelis, 9 October 1759, Göttingen University Archive Cod. MS. Mich. 320, fols 618–19 (transcription by Albert Kreyer; translation by Wolfgang Köberer).
37. *London Chronicle*, 24–26 August 1762.
38. Nevil Maskelyne to Edmund Maskelyne, 29 December 1763, NMM REG09/000037 (CDL, <http://cudl.lib.cam.ac.uk/view/MS-REG-00009-00037/316> [accessed 3 December 2013]).
39. 'Harrison Journal', 1817, State Library of Victoria H17809, p. 112.
40. BoL, confirmed minutes, 9 February 1765, CUL RGO 14/5, p. 77 (CDL, <http://cudl.lib.cam.ac.uk/view/MS-RGO-00014-00005/81> [accessed 3 December 2013]).
41. Ibid., p. 78 (CDL, <http://cudl.lib.cam.ac.uk/view/MS-RGO-00014-00005/82> [accessed 3 December 2013]).

CHAPTER 4

Fig. 1

The Royal Observatory from the south-east, unknown artist, c.1770 (NMM AST0042)

Fig. 2

The Nautical Almanac and Astronomical Ephemeris for 1767 (NMM/ZBA5690)

Fig. 3

Tables Requisite to be used with the Astronomical and Nautical Almanac (London, 1766) (NMM PBH6275)

Fig. 4

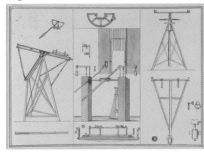

Instruments at the Royal Observatory, Greenwich, drawing by John Charnock, about 1785 (NMMPAF2956)

Fig. 5

Nevil Maskelyne's 'observing suit', about 1765 (NMM ZBA4675–6)

Fig. 6

Nevil Maskelyne, by Louis François Gérard van der Puyl, 1785 (The Royal Society P/0088)

Fig. 7

List of reference books required by the computers, compiled by Nevil Maskelyne in 1799 (CUL RGO 4/324, f. 21v)

Fig. 8

List by Nevil Maskelyne of work allocated to computers and comparers in 1791–93 for the 1803 *Nautical Almanac* (CUL RGO 4/324, f. 4r)

Fig. 9

Date and signature on H4 (NMM/MoD ZAA0037)

Fig. 10

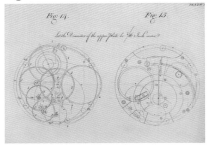

Drawings from *The Principles of Mr. Harrison's Timekeeper* (London, 1767) (NMM ZAA0643)

Fig. 11

Marine timekeeper K1, by Larcum Kendall, 1769 (NMM/MoD ZAA0038)

Fig. 12

Marine timekeepers, by John Arnold, c.1771 (The Royal Society T/002, T/003)

Fig. 13

Horloge marine no. 8, by Ferdinand Berthoud, 1767 (CNAM Paris 01389)

Fig. 14

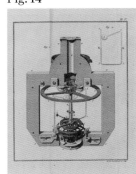

Le Tellier, *Journal du Voyage de M. le Marquis de Courtanvaux* (Paris, 1768) (NMM PBA1788)

Fig. 15

Marine timekeeper, by William Snellen, c.1775 (NMM ZAA0282)

Fig. 16

Marine timekeeper H5, by John Harrison, completed 1770 (The Worshipful Company of Clockmakers)

233

Fig. 17

Concluding page from H5's Kew Observatory trial, by Stephen Demainbray, 1772 (King's College London K/MUS 1/6)

Fig. 18

Medallion portrait of John Harrison by James Tassie, c.1776 (NMM OBJ0563)

Fig. 19

Printed lunar-distance form, published by Robert Bishop, 1768 (NMM G298:1/3)

1. Foreword by the Commissioners of Longitude, *Nautical Almanac and Astronomical Ephemeris for the Year 1767* (London, 1766) [no page number].
2. *An Act for explaining and rendering more effectual two Acts ...* (5 Geo. 3, c. 20), 1765.
3. Anonymous letter, *The Gentleman's Magazine*, 35 (1765), 34–35.
4. BoL confirmed minutes, 9 February 1765, CUL RGO 14/5, p. 77 (CDL, <http://cudl.lib.cam.ac.uk/view/MS-RGO-00014-00005/81> [accessed 3 December 2013]).
5. Nevil Maskelyne to Edmund Maskelyne, 15 May 1766, NMM REG 09/000037.
6. Ibid.
7. 'Qualities to be required for an Assistant, May 19. 1787', Memorandum Book 1782–88, Story Maskelyne of Purton papers, Wiltshire and Swindon Archives 1390/2d.
8. BoL confirmed minutes, 9 February 1765, CUL RGO 14/5, p. 78 (CDL, <http://cudl.lib.cam.ac.uk/view/MS-RGO-00014-00005/82> [accessed 3 December 2013]).
9. Harrison Journal, 1817, State Library of Victoria H17809, p. 55.
10. Nevil Maskelyne, *An Account of the Going of Mr. John Harrison's Watch, at the Royal Observatory* (London, 1767), p. 24.
11. Letter to Edmund Maskelyne, 15 May 1766.
12. BoL confirmed minutes, 26 April 1766, CUL RGO 14/5, p. 124 (CDL, <http://cudl.lib.cam.ac.uk/view/MS-RGO-00014-00005/128> [accessed 3 December 2013]).
13. Copy of Kendall to the Commissioners of Longitude, 26 April 1766, Kendall Papers, BL Add MS 39822, f. 40.
14. William Ludlum to Larcum Kendall, 24 October 1765, Kendall Papers, BL Add MS 39822, ff. 36–7.
15. Extract from Minutes made at a Board of Longitude meeting, 12 September 1765, Kendall Papers, BL Add MS 39822, f. 35.
16. Copy of Kendall to the Commissioners of Longitude, 26 May 1770, Kendall Papers, BL Add MS 39822, f. 47.
17. Kendall to the Commissioners of Longitude, 26 May 1770.
18. Copy of Kendall to the Commissioners of Longitude, 13 June 1772, Kendall Papers, BL Add MS 39822, f. 48.
19. BoL confirmed minutes, 14 March 1767, CUL RGO 14/5, p. 146 (CDL, <http://cudl.lib.cam.ac.uk/view/MS-RGO-00014-00005/150> [accessed 3 December 2013]).
20. Quoted in A. J. Turner, 'Berthoud in England, Harrison in France: The Transmission of Horological Knowledge in 18th Century Europe', *Antiquarian Horology*, 20 (1992), 219–39 (p. 233).
21. Johan Horrins [pseudonym of John Harrison's grandson, John Harrison], *Memoirs of a Trait in the Character of George III* (London, 1835), p. 6.
22. BoL confirmed minutes, 28 November 1772, CUL RGO 14/5, p. 231 (CDL <http://cudl.lib.cam.ac.uk/view/MS-RGO-00014-00005/235> [accessed 3 December 2013]).
23. Quoted in Humphrey Quill, *John Harrison: The Man Who Found Longitude* (London, 1966), p. 196.
24. Quoted in Ibid., p. 198–99.
25. BoL confirmed minutes, 24 April 1773, CUL RGO 14/5, p. 240 (CDL, <http://cudl.lib.cam.ac.uk/view/MS-RGO-00014-00005/244> [accessed 3 December 2013]).
26. *Parliamentary History of the House of Commons*, 17 (1813), cols. 812–3, quoted in Derek Howse, *Nevil Maskelyne – The Seaman's Astronomer* (Cambridge, 1989), p. 125.
27. *House of Commons Journal*, 34 (1772–74), 367, quoted in Howse, *The Seaman's Astronomer* (1989), p. 125.
28. BoL confirmed minutes, 2 March 1771, CUL RGO 14/5, p. 203 (CDL, <http://cudl.lib.cam.ac.uk/view/MS-RGO-00014-00005/207> [accessed 3 December 2013]).
29. W. Emerson, *The Mathematical Principles of Geography, Navigation and Dialling* (London, 1770), p. 172.

CHAPTER 5

Fig. 1

'Various articles at Nootka Sound', by John Webber
(NMM PAI3913)

Fig. 2

Captain James Cook, by Nathaniel Dance, 1775–76
(NMM BHC2628)

Fig. 3

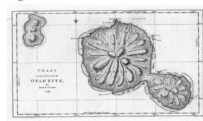

'Chart of the Island of Otaheite [Tahiti]' by James Cook, 1769
(NMM G267:47/1)

Fig. 4

Cook's *Resolution* in the Marquesas Islands, by William Hodges, 1774
(NMM PAF5791)

Fig. 5

Captain Cook's journal on the *Resolution*, 7 January 1774
(NMM JOD/20)

Fig. 6

View of Point Venus and Matavai Bay, Tahiti, by William Hodges, 1773
(NMM/MoD BHC1937)

Fig. 7

'View of Maitavie [Matavai] Bay', by William Hodges, 1776
(NMM/MoD BHC1932)

Fig. 8

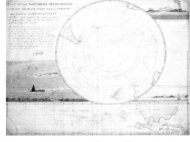

'Part of the Southern Hemisphere shewing the Resolution's Track through the Pacific and Southern Ocean', by Joseph Gilbert, *c.*1775
(UK Hydrographic Office)

Fig. 9

Draft list of instruments, by Nevil Maskelyne, dated 22 May 1776
(NMM AGC/8/29)

Fig. 10

'*Resolution* and *Discovery* in Ship Cove, Nootka Sound', by John Webber, 1778
(NMM PAJ2959)

Fig. 11

Chart of New Zealand, by James Cook, 1772
(NMM G263:1/2)

Fig. 12

Sextant, by Ramsden, London, c.1772
(NMM NAV1236)

Fig. 13

'A Sea Otter', by John Webber
(NMM PAI3916)

Fig. 14

Marine timekeeper K2, by Larcum Kendall, London, 1771
(NMM ZAA0078)

Fig. 15

A branch of the breadfruit tree, from John Hawkesworth, *An Account of the Voyages Undertaken ... in the Southern Hemisphere* (London, 1773)
(NMM PAJ1965)

Fig. 16

'The Mutineers turning Lt Bligh and part of the officers and crew adrift', by Robert Dodd, 1790
(NMM PAH9205)

Fig. 17

Marine timekeeper K3, by Larcum Kendall, London, completed in 1774
(NMM/MoD ZAA0111)

Fig. 18

'A Chart showing part of the Coast of N.W. America', drawn by George Vancouver, published by Robinson and Edwards, 1798
(NMM G278:1/1)

Fig. 19

'Wreck Reef Bank', by William Westall, August 1803
(MoD/NMM BHC1164)

Fig. 20

Lapérouse's officers with the islanders on Sakhalin, from *The Voyage of La Pérouse Round the World* (London, 1798)
(NMM PBN1856/2)

1. Christine Holmes (ed.), *Captain Cook's Second Voyage: The Journals of Lieutenants Elliott and Pickersgill* (London, 1984), p. 45.
2. Quoted in Randolph Cock, 'Precursors of Cook: The Voyages of the *Dolphin*, 1764–8', *Mariner's Mirror*, 85 (1999), 30–52 (p. 46).
3. Quoted in J. C. Beaglehole, *The Life of Captain James Cook* (London, 1974), p. 154.
4. J. C. Beaglehole (ed.), *The Journals of Captain James Cook on his Voyages of Discovery: The Voyage of the Endeavour 1768–1771* (Cambridge, 1955), pp. 52–53.
5. Beaglehole, *Endeavour*, p. 392.
6. BoL, confirmed minutes, 28 November 1771, CUL RGO 14/5, p. 207 (CDL, <http://cudl.lib.cam.ac.uk/view/MS-RGO-00014-00005/211> [accessed 3 December 2013]).
7. BoL, confirmed minutes, 14 May 1772, CUL RGO 14/5, pp. 223–24 (CDL, <http://cudl.lib.cam.ac.uk/view/MS-RGO-00014-00005/227> [accessed 3 December 2013]).
8. J. C. Beaglehole (ed.), *The Journals of Captain James Cook on his Voyages of Discovery – II: The Voyage of the Resolution and Adventure 1772–1775* (Cambridge, 1969), p. 445.
9. Beaglehole, *Resolution and Adventure*, p. cxii.
10. Quoted in Derek Howse, 'Captain Cook's Marine Timekeepers – I: The Kendall Watches', *Antiquarian Horology*, 6 (1969), 190–205 (p. 194).
11. Beaglehole, *Resolution and Adventure*, p. 660.
12. Ibid., pp. clxviii–clxix.
13. BoL instructions issued to William Wales, in Beaglehole, *Resolution and Adventure*, p. 726.
14. J. C. Beaglehole (ed.), *The Journals of Captain James Cook on his Voyages of Discovery – III: The Voyage of the Resolution and Discovery, 1776–1780 Part I* (1967, reprinted Woodbridge, 1999), pp. 454–55.
15. James King, log of the *Resolution*, 12 February 1776 – 2 February 1778, TNA ADM 55/116, fol. 3r.
16. Beaglehole, *Resolution and Adventure*, pp. 524–55.
17. Ibid., pp. 579–80.
18. Beaglehole, *Resolution and Discovery*, p. 685.
19. King, log of the *Resolution*, fol. 2r.
20. Beaglehole, *Resolution and Adventure*, p. 726.
21. Ibid., pp. 524–25.
22. Christopher Lloyd and R. C. Anderson (eds), *A Memoir of James Trevenen*, Navy Records Society, 101 (1959), 53–54.
23. Georg Forster, *Cook the Discoverer* (Sydney, 2007), p. 197.
24. William Bligh, *A Voyage to the South Sea* (London, 1792), p. 5.
25. William Bligh, *Narrative of the Mutiny, on board His Majesty's Ship Bounty* (London, 1790), p. 3.
26. John Hawkesworth, *An Account of the Voyages Undertaken … in the Southern Hemisphere*, 3 vols (London, 1773), I, p. 341.
27. George Vancouver, *A Voyage of Discovery to the North Pacific Ocean and Round the World 1791–1795*, ed. by W. Kaye Lamb, 4 vols (London, 1984), I, p. 309.
28. Ibid., pp. 313, 317, 320.
29. Ibid., p. 320.
30. James Smedley to the Commissioners of Longitude on behalf of John Crosley, 29 November 1803, CUL RGO 14/1, fols 171–72 (CDL, <http://cudl.lib.cam.ac.uk/view/MS-RGO-00014-00001/335> [accessed 3 December 2013]).
31. Matthew Flinders to Nevil Maskelyne, 25 May 1802, CUL RGO 14/68, fol. 17v (CDL, <http://cudl.lib.cam.ac.uk/view/MS-RGO-00014-00068/34> [accessed 3 December 2013]).
32. Louis de Bougainville, *A Voyage Round the World* (London, 1772), p. 242.
33. Luis R. Martínez-Cañavate (ed.), *La Expedición Malaspina*, 6 vols (Barcelona, 1994), VI, p. 124 (translation by Juan Pimentel).
34. Peter Fidler, journal entry, 18 December 1791, HBCA E 3/1, fol. 37, quoted in Peter Broughton, 'The Accuracy and Use of Sextants and Watches in Rupert's Land in the 1790s', *Annals of Science*, 66 (2009), 200–29 (p. 214).
35. Thomas Jefferson, Instructions to Captain Merriweather Lewis, 20 June 1803, in Donald Jackson (ed.), *Letters of the Lewis and Clark Expedition with Related Documents, 1783–1854* (University of Illinois, 1978), p. 61.
36. Robert Patterson to Thomas Jefferson, 15 March 1803, in Jackson, *Letters of the Lewis and Clark Expedition*, p. 29.
37. Thomas Brisbane, speech at Glasgow Observatory, 16 December 1836, State Library of New South Wales Mitchell MSS 1191/1/521.
38. Matthew Flinders, 'Biographical tribute to his cat Trim 1809', NMM FLI/11, p. 1.
39. William Wales, *The Method of Finding the Longitude at Sea by Time-Keepers* (London, 1794), pp. iv–v.

CHAPTER 6

Fig. 1

'March of Intellect – Lord, how this world improves as we grow older', by William Heath, published by Thomas McLean, London, 1829 (Private collection)

Fig. 2

Thomas Mudge, by Nathaniel Dance, c.1772 (Science Museum 1985-1362)

Fig. 3

Marine timekeeper 'Green', by Thomas Mudge, 1777 (Private collection)

Fig. 4

Mudge-type marine timekeeper no. 4, by Howells and Pennington, London, c.1794, in a later box (NMM/MoD ZAA0133)

Fig. 5

John Arnold and family, by Robert Davy, c.1783 (Science Museum 1868-248)

Fig. 6

Pocket chronometer no. 36, by John Arnold, London, 1778 (NMM ZBA1227)

Fig. 7

Thomas Earnshaw, by Martin Archer Schee, c.1808 (NMM BHC2674)

Fig. 8

Marine chronometer, by John Arnold, London, c.1790 (NMM ZAA0012)

Fig. 9

Marine chronometers nos. 512 and 524, by Thomas Earnshaw, London, c.1800 (NMM ZAA0006, ZAA0732)

Fig. 10

Escapement model, by Thomas Earnshaw, 1804 (NMM/MoD ZAA0123)

Fig. 11

Jesse Ramsden, by Robert Home, c.1791 (The Royal Society P/0107)

Fig. 12

Ramsden's second dividing engine, from *Description of an Engine for Dividing Mathematical Instruments* (London, 1777) (NMM PBG0968)

Fig. 13

Sextant by Nathaniel Worthington, London, c.1840 (NMM NAV1214)

Fig. 14

'Little Midshipman' trade sign, late eighteenth century
(NMM AAB0173)

Fig. 15

Models of a scoring machine and a mortising machine, by Marc Isambard Brunel and Henry Maudslay, c.1803
(NMM MDL0013, MDL0016)

Fig. 16

Babbage's Difference Engine, from *Harper's New Monthly Magazine*, 1865
(Private collection)

Fig. 17

Four of the twenty-one volumes of Babbage's *Specimen of Logarithmic Tables* (London, 1831)
(Royal Observatory, Edinburgh)

Fig. 18

Mechanical log, by Edward Massey, London, c.1830
(NMM NAV0728)

Fig. 19

William Chavasse's proposed observing platform, submitted in 1813
(CUL RGO 14/36, fol. 51)

Fig. 20

Samuel Parlour's shoulder-mounted apparatus, submitted in 1824
(CUL RGO 14/30, fol. 504)

Fig. 21

Ralph Walker, published by James Asperne, 1803
(NMM PAD3061)

Fig. 22

Azimuth compass, designed by Ralph Walker, London, c.1793
(NMM NAV0263)

Fig. 23

Mercury log glass, made by William and Thomas Gilbert, London, c.1817
(NMM NAV0744)

Fig. 24

Insulating compass, by Jennings and Company, London, c.1818
(NMM ACO1517)

Fig. 25

John Couch's 'Calitsa' for riding through surf, 1819
(CUL RGO 14/44, fol. 90)

1. C. B. Tennyson (ed.), *A Carlyle Reader* (Cambridge, 1984), p. 34.
2. James Boswell, *The Life of Samuel Johnson* (London, 1830), p. 535.
3. Daniel Defoe, *A Plan of English Commerce* (London, 1728), p. 300.
4. Sir Joseph Banks to William Windham MP, 20 April 1793, in Neil Chambers (ed.), *The Letters of Sir Joseph Banks: A Selection, 1768–1820* (London, 2000), p. 151.
5. Quoted in Jonathan Betts, 'Arnold, John (1735/6–1799)', *Oxford Dictionary of National Biography* (Oxford, 2004; online edn, May 2009), <http://www.oxforddnb.com/view/article/677> [accessed 6 July 2013].
6. William Ludlam to Nevil Maskelyne, 23 October 1783, quoted in *Report from the Select Committee of the House of Commons ... to whom the Petition of Thomas Mudge, Watchmaker, was referred* (London, 1793), p. 96.
7. 'Dr Maskelyne's Abstract of the going of several time-keepers in a voyage to the Cape of good Hope', CUL RGO 14/25, fol. 184v (CDL, <http://cudl.lib.cam.ac.uk/view/MS-RGO-00014-00025/375> [accessed 3 December 2013]).
8. Thomas Earnshaw, *Explanation of Timekeepers constructed by Thomas Earnshaw* (London, 1805), p. 9.
9. *The Morning Chronicle*, 4 February 1806, p. 1.
10. Abraham Rees, *The Cyclopædia; or Universal Dictionary of Arts, Sciences, and Literature*, VIII (London, 1819), 8.
11. Quoted in Anita McConnell, *Jesse Ramsden (1735–1800): London's Leading Scientific Instrument Maker* (Aldershot, 2007), p. 48.
12. Charles Dickens, *Dombey and Son* (London, 1848), p. 24.
13. Charles Babbage, *On the Economy of Machinery and Manufactures* (London, 1832), p. 39.
14. H. W. Buxton, *Memoir of the Life and Labours of the Late Charles Babbage*, ed. by Anthony Hyman (Cambridge, Mass., 1988), p. 46.
15. Quoted in Peter J. Turvey, 'Sir John Herschel and the Abandonment of Charles Babbage's Difference Engine No. 1', *Notes and Records of the Royal Society of London*, 45 (1991), 165–76 (p. 173).
16. G. S. Hillard (ed.), *Life, Letters, and Journals of George Ticknor* (London, 1876).
17. Quoted in Doron Swade, 'Babbage, Charles (1791–1871)', *Oxford Dictionary of National Biography* (Oxford, 2004; online edn, May 2009), <http://www.oxforddnb.com/view/article/962> [accessed 6 July 2013].
18. Bartholomew de Sanctis to the Board of Longitude, 26 October 1823, CUL RGO 14/40, fol. 431r (CDL, <http://cudl.lib.cam.ac.uk/view/MS-RGO-00014-00040/303> [accessed 3 December 2013]).
19. Francis Higginson, 'Instruments for facilitating and rendering accurate the methods of finding the Latitude and Longitude at Sea', 1828, CUL RGO 14/31, fol. 278r (CDL, <http://cudl.lib.cam.ac.uk/view/MS-RGO-00014-00031/609> [accessed 3 December 2013]).
20. John Ross, *A Voyage of Discovery made under the orders of the Admiralty, in His Majesty's ships Isabella and Alexander* (London, 1819), Appendix, p. cxxv.
21. H. C. Jennings to Lt General W. Wyngard, 4 December 1817, CUL RGO 14/31, fol. 214 (CDL, <http://cudl.lib.cam.ac.uk/view/MS-RGO-00014-00031/483> [accessed 3 December 2013]).
22. Lieutenant John Couch to BoL, 1818–23, CUL RGO 14/44, fols 80r–94v (CDL, <http://cudl.lib.cam.ac.uk/view/MS-RGO-00014-00044/171> [accessed 3 December 2013]).
23. John Bradley to the Lords of the Royal Navy, 9 January 1787, CUL RGO 14/39, fol. 29r (CDL, <http://cudl.lib.cam.ac.uk/view/MS-RGO-00014-00039/57> [accessed 3 December 2013]).
24. Henry Croaker to BoL, 1819–20, CUL RGO 14/39, fols 74r–111v (CDL, <http://cudl.lib.cam.ac.uk/view/MS-RGO-00014-00039/167> [accessed 3 December 2013]).
25. Walter Bedford to BoL, 7 February 1783, CUL RGO 14/39, fol. 17r (CDL, <http://cudl.lib.cam.ac.uk/view/MS-RGO-00014-00039/35> [accessed 3 December 2013]).
26. John Croker in a House of Commons debate on the Corporate Funds Bill, 4 July 1828, quoted in *The Times*, 5 July 1828.

CHAPTER 7

Fig. 1

'Loss of the *Magnificent*, 25 March 1804', by John Christian Schetky, 1839
(NMM BHC0534)

Fig. 2

The time ball at the Royal Observatory, Greenwich, *Illustrated London News*, 9 November 1844
(NMM C4102)

Fig. 3

Deck scene, with two men taking Sun observations from the quarterdeck, by Thomas Streatfeild, 1820
(NMM PAI4318)

Fig. 4

'Life on the ocean, representing the usual occupations of the young officers in the steerage of a British frigate at sea', by Augustus Earle, c.1820–37
(NMM BHC1118)

Fig. 5

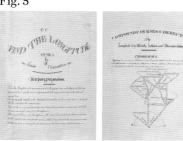

Navigational workbook compiled by John Marshall, 1810
(NMM NVT/4)

Fig. 6

'Landing the Treasures, or Results of the Polar Expedition!!!', by George Cruikshank, 1819
(NMM PAG8511)

Fig. 7

'Discoveries of Capts Ross, Parry & Franklin in the Arctic regions from the year 1818 to 1827', from John Thomson's *New General Atlas* (London, c.1830)
(NMM G285:1/2)

Fig. 8

Commander James Clark Ross, by John R. Wildman, 1834
(NMM BHC2981)

Fig. 9

Marine chronometer, by George Margetts, c.1790
(NMM G241:4/38)

Fig. 10

'West Coast of Africa, Sheet VII, ... surveyed by Captain W. F. W. Owen, Commander E. Belcher, and Lieutenant W. Arlett', 1836
(NMM ZBA0672)

Fig. 11

Advertisement for Price's candles, c.1850
(Lambeth Archives #4248)

Fig. 12

The *Beagle* in the Beagle Channel, Tierra del Fuego, by Conrad Martens, c.1834
(NMM PAF6228)

241

Fig. 13

The south-eastern part of Tierra del Fuego, 1877 (NMM STK244:2/2(3))

Fig. 14

Portrait Cove, Beagle Channel, by Conrad Martens, *c.*1834 (NMM PAF6242)

Fig. 15

Parallel rule, made by Nathaniel Worthington, *c.*1836 (NMM NAV0605)

Fig. 16

The Victoria River, 1845 (NMM STK262:8/3)

Fig. 17

Theodolite by J. Dollond & Son, *c.*1840, and sextant by Worthington & Allen, *c.*1831 (NMM NAV1461, NAV1174)

Fig. 18

Marine chronometer no. 294, by John Roger Arnold, *c.*1807 (NMM ZAA0292)

Fig. 19

'Chart of the World, showing Tracks followed by Sailing and Low Powered Steam Vessels', 1888 (NMM G201:1/62)

Fig. 20

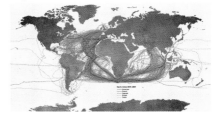

Visualization of international shipping routes, based on the Climatological Database for the World's Oceans 1800–1850 (CLIWOC)

PROLOGUE

page 10

A map of the world, by Paolo Forlani, published by Fernando Bertelli, 1565 (NMM G201.1/35)

EPILOGUE

page 224

'The *Great Western* riding a tidal wave, 11 December 1844', by Joseph Walter, 1845 (NMM BHC2379)

1. 'The Royal Observatory, Greenwich', *The Weekly Visitor*, 24 February 1833.
2. Richard Owen, 'Essay on Chronometers', in W. F. W. Owen, *Tables of Latitudes and Longitudes by Chronometer* (London, 1827), pp. 1–34 (p. 3).
3. *Nautical Magazine*, 28 October 1833, p. 680, quoted in Howse, *Greenwich Time*, p. 83.
4. Reproduced in H. W. Dickinson, *Educating the Royal Navy: Eighteenth- and Nineteenth-Century Education for Officers* (London and New York, 2007), p. 14.
5. Quoted in Christopher Lloyd, 'The Royal Naval Colleges at Portsmouth and Greenwich', *Mariner's Mirror*, 52 (1966), 145–65 (p. 148).
6. Captain Montagu Burrows, quoted in Lloyd, 'The Royal Naval Colleges', p. 148.
7. John Croker, on the 'Longitude Discovery Bill', 6 March 1818, *Hansard*, 37 (1818), cc. 876–80, col. 876.
8. John Barrow, *A Chronological History of the Voyages into the Arctic Regions* (London, 1818), pp. 364–65.
9. Barrow, *A Chronological History*, pp. 368–69.
10. W. E. Parry, *Journal of a Voyage for the Discovery of a North-West Passage from the Atlantic to the Pacific* (London, 1821), p. 72.
11. W. F. W. Owen, *A Narrative of a Voyage to Explore the Shores of Africa, Arabia and Madagascar* (London, 1833), p. 8.
12. Ibid., pp. 7–8.
13. Ibid., p. 7.
14. Ibid.
15. Richard Owen, 'Essay on Chronometers', p. 3.
16. Ibid.
17. Quoted in Jordan Goodman, 'The Hell-Borne Traffic: William Owen and the African Slave Trade', *Geographical*, 79 (2007), 63–65 (p. 64).
18. Philip Parker King et al., *Narrative of the Surveying Voyages of His Majesty's Ships Adventure and Beagle ...*, 4 vols (London, 1839), I, p. xvii.
19. 'South America West Coast Original Directions by Captain FitzRoy HMS Beagle, 1834–5', UKHO OD44 SA.2.6.
20. John Lort Stokes, *Discoveries in Australia, with an Account of the Coasts and Rivers Explored and Surveyed During the Voyage of H.M.S. Beagle, in the Years 1837-38-39-40-41-42-43*, 3 vols (London, 1846), II, pp. 526–27.
21. Robert FitzRoy to Francis Beaufort, 10 May 1833, UKHO, in R. D. Keynes (ed.), *The Beagle Record: Selections from the Original Pictorial Records and Written Accounts of the Voyage of H.M.S. Beagle* (Cambridge, London, New York and Melbourne, 1979), p. 131.
22. Robert FitzRoy, 'Appendix' to King et al., *Narrative of the Surveying Voyages*, II, p. 331.
23. FitzRoy to Beaufort, 26 October 1833, UKHO, in Keynes (ed.), *The Beagle Record*, p. 162.
24. Ibid., p. 163.
25. 'South America West Coast Original Directions by Captain FitzRoy HMS Beagle, 1834–5', UKHO OD44 SA.2.6.
26. Ibid.
27. Ibid.
28. Stokes, *Discoveries in Australia*, II, pp. 359–60.
29. Ibid.
30. Francis Beaufort to John Barrow, 20 December 1836, UKHO MB2, pp. 345–46.
31. John F. W. Herschel, 'Address of the President', in *Report of the Fifteenth Meeting of the British Association for the Advancement of Science* (London, 1846), pp. xxvii–xliv, xxxiv, xxix.

PROLOGUE

1. *An Act for providing a Publick Reward for such Person or Persons as shall discover the Longitude at Sea* (13 Anne, c. 14), 1713.

EPILOGUE

1. George Gordon Byron, *The Works of Lord Byron, Complete in One Volume* (London, 1837), p. 7.

BIBLIOGRAPHY

Primary Sources

The archives of the Board of Longitude are held in the Royal Greenwich Observatory manuscripts at Cambridge University Library. Along with papers of the Astronomers Royal, Nevil Maskelyne and John Pond, from Cambridge and the National Maritime Museum, they have been digitized and are freely available as part of the Cambridge Digital Library at: http://cudl.lib.cam.ac.uk/collections/longitude

References to manuscript and contemporary printed sources, most of which can be found at Eighteenth Century Collections Online, are given in the endnotes. The following abbreviations have been used:

BL	British Library
BoL	Board of Longitude
CDL	Cambridge Digital Library
CUL	Cambridge University Library
HBCA	Hudson's Bay Company Archives, Manitoba
NMM	National Maritime Museum, Greenwich
RGO	Royal Greenwich Observatory Archives
TNA	The National Archives of the UK
UKHO	United Kingdom Hydrographic Office

Further Reading

General Background

The longitude story, particularly its British context, is perhaps best introduced in Derek Howse, *Greenwich Time and the Longitude* (London, 1997). Dava Sobel's bestselling *Longitude: The True Story of a Lone Genius Who Solved the Greatest Scientific Problem of His Time* (London, 1996) provides a quick read that gives the story from John Harrison's perspective. Deriving from the conference that inspired Sobel's work, W. J. H. Andrewes (ed.), *The Quest for Longitude* (Harvard, 1996) has a strong focus on horological developments but also contains useful essays on the international background and some of the other methods for determining longitude. The French side of the story is told in more detail in F. Marguet, *Histoire Générale de la Navigation du XVe au XXe Siècle* (Paris, 1931) and in the essays in Vincent Jullien (ed.), *Le Calcul des Longitudes* (Rennes, 2002).

John Harrison's life and work have been dealt with in full in Humphrey Quill, *John Harrison: The Man Who Found Longitude* (London, 1966), and more briefly in Jonathan Betts, *Harrison* (London, 2007). Harrison's work is put into its horological context in Rupert Gould's still largely authoritative *The Marine Chronometer: Its History and Development* (London, 1923) and Jonathan Betts, *Marine Chronometers at Greenwich* (Oxford, forthcoming). David S. Landes, *Revolution in Time: Clocks and the Making of the Modern World* (2nd edition, London, 2000) and Paul Glennie and Nigel Thrift, *Shaping the Day: A History of Timekeeping in England and Wales, 1300–1800* (Oxford, 2009) have also discussed how marine timekeepers fit into a longer economic, industrial and social history. For more detail on the economic background, see Joel Mokyr, *The Enlightened Economy: An Economic History of Britain, 1700–1850* (New Haven and London, 2009).

A good overview of the history of the Board of Longitude can be found in Peter Johnson, 'The Board of Longitude 1714–1828', *Journal of the British Astronomical Association*, 99 (1989), 63–66. Derek Howse wrote a biography of one of the most significant figures in the Board's history, in *Nevil Maskelyne: The Seaman's Astronomer* (Cambridge, 1989). The history of the Royal Observatory, where Maskelyne and the other Astronomers Royal worked, is told in Eric G. Forbes, A. J. Meadows and Derek Howse, *Greenwich Observatory: The Royal Observatory at Greenwich and Herstmonceux, 1675–1975*, 3 vols (London, 1975).

Charles H. Cotter, *A History of Nautical Astronomy* (London, 1968) provides a comprehensive account of celestial navigation, while Peter Ifland, *Taking the Stars: Celestial Navigation from Argonauts to Astronauts* (Malabar, 1998) and J. A. Bennett, *The Divided Circle: A History of Instruments for Astronomy, Navigation and Surveying* (Oxford, 1987) concentrate on the development of the instruments. The history of navigation more generally is covered in E. G. R. Taylor, *The Haven-Finding Art* (London, 1956), W. E. May, *A History of Marine Navigation* (Guildford, 1973) and Donald S. Johnson and Juha Nurminen, *The History of Seafaring* (London, 2007), among others.

Many of the essays in John B. Hattendorf, *The Oxford Encyclopedia of Maritime History* (Oxford, 2007) provide excellent starting points for a range of relevant topics. The definitive works on British naval history more specifically are N. A. M. Rodger's *The Wooden World* (London, 1986) and *The Command of the Ocean: A Naval History of Britain, 1649–1815* (London, 2004). For the history of exploration, many useful pieces can be found in Derek Howse (ed.), *Background to Discovery: Pacific Exploration from Dampier to Cook* (Berkeley, 1990) and David Mackay, *In the Wake of Cook* (Beckenham, 1985), while Glyn Williams, *Arctic Labyrinths: The Quest for the Northwest Passage* (London, 2010) offers a very readable narrative of maritime exploration in one part of the world.

The relationship between navy and science that developed in the eighteenth century is discussed in John Gascoigne, *Science in the Service of Empire: Joseph Banks, the British State and the Uses of Science in the Age of Revolution* (Cambridge, 1998). Banks and his context also form a core thread in Richard Holmes's *Age of Wonder: How the Romantic Generation Discovered the Beauty and Terror of Science* (London, 2009). For a more general background, see Roy Porter (ed.), *The Cambridge History of Science Vol. 4: Eighteenth-Century Science* (Cambridge, 2003).

An essential source for most of the British individuals mentioned in this book is the *Oxford Dictionary of National Biography* (Oxford, 2004; online edition, Jan 2008). Additional works that were particularly useful in the writing of individual chapters are listed below.

Chapter 1

H. V. Bowen, John McAleer, Robert J. Blyth, *Monsoon Traders: The Maritime World of the East India Company* (London, 2011)

K. N. Chaudhuri, *The Trading World of Asia and the English East India Company, 1660–1760* (Cambridge, 1978)

Mike Dash, *Batavias's Graveyard* (London, 2002)

Gillian Hutchinson, *Medieval Ships and Shipping* (Leicester, 1994)

María M. Portuondo, *Secret Science: Spanish Cosmography and the New World* (Chicago & London, 2009)

Diana and Michael Preston, *A Pirate of Exquisite Mind. The Life of William Dampier: Explorer, Naturalist and Buccaneer* (London, 2004)

Rachel Souhami, *Selkirk's Island* (London, 2002)

David W. Waters, *The Art of Navigation in England in Elizabethan and Early Stuart Times* (London, 1958)

Chapter 2

D. J. Bryden, 'Magnetic inclinatory needles: Approved by the Royal Society?', *Notes and Records of the Royal Society of London*, 47 (1993), 17–31

Karel Davids, 'Dutch and Spanish global networks of knowledge in the early modern period: Structures, connections, changes', in Lissa Roberts (ed.), *Centres and Cycles of Accumulation in and around the Netherlands during the Early Modern Period* (Münster, 2011), 29–52

Lisa Jardine, *Going Dutch: How England Plundered Holland's Glory* (London, 2008)

Lisa Jardine, 'Scientists, sea trials and international espionage: Who really invented the balance-spring watch?', *Antiquarian Horology*, 9 (2006), 663–83

A. R. T. Jonkers, *Earth's Magnetism in the Age of Sail* (Baltimore, 2003)

Kevin C. Knox and Richard Noakes, *From Newton to Hawking: A History of Cambridge University's Lucasian Professors of Mathematics* (Cambridge, 2003)

Stephen Pumfrey, *Latitude and the Magnetic Earth* (London, 2002)

Simon Werrett, *Fireworks: Pyrotechnic Arts and Sciences in European History* (Chicago and London, 2010)

Richard S. Westfall, *Never at Rest: A Biography of Isaac Newton* (Cambridge, 1983)

Frances Willmoth (ed.), *Flamsteed's Stars: New Perspectives on the Life of the First Astronomer Royal, 1646–1719* (Bristol, 1997)

Chapter 3

Daniel A. Baugh, 'The Sea-Trial of John Harrison's Chronometer, 1736', *Mariner's Mirror*, 64 (1978), 235–40

Silvio A. Bedini, *Thinkers and Tinkers: Early American Men of Science* (New York, 1975)

J. A. Bennett 'Catadioptrics and Commerce in Eighteenth-Century London', *History of Science*, 44 (2006), 247–87

J. G. Burke (ed.), *The Uses of Science in the Age of Newton* (Berkeley, 1983)

Markman Ellis, *The Coffee House* (London, 2004)

Alan Ereira, 'The Voyages of H1', *Mariner's Mirror*, 87 (2001), 144–49

E. G. Forbes, *The Birth of Navigational Science* (London, 1974)

E. G. Forbes, *Tobias Mayer (1723–62): Pioneer of Enlightened Science in Germany* (Göttingen, 1980)

Larry Stewart, *The Rise of Public Science* (Cambridge, 1992)

Jeffrey R. Wigelsworth, *Selling Science in the Age of Newton: Advertising and the Commoditization of Knowledge* (Farnham, 2010)

Glyn Williams, *The Prize of All the Oceans* (London, 1999)

Chapter 4

J. A. Bennett, 'The Travels and Trials of Mr Harrison's Timekeeper', in M.N. Bourguet, C. Licoppe and H. O. Sibum (eds), *Instruments, Travel and Science: Itineraries of Precision from the Seventeenth to the Twentieth Century* (London, 2002), 75–95

Mary Croarken, 'Tabulating the Heavens: Computing the Nautical Almanac in 18th-Century England', *IEEE Annals of the History of Computing*, 25:3 (2003), 48–61

Alan Morton and Jane Wess, *Public and Private Science: The King George III Collection* (Oxford, 1993)

D. H. Sadler, *Man is Not Lost: A Record of Two Hundred Years of Astronomical Navigation with the Nautical Almanac, 1767–1967* (London, 1968)

A. J. Turner, 'Berthoud in England, Harrison in France: The transmission of horological knowledge in 18th century Europe', *Antiquarian Horology*, 20 (1992), 219–39

Chapter 5

John Bach (ed.), *The Bligh Notebook* (Sydney, 1987)

J. C. Beaglehole (ed.), *The Journals of Captain James Cook on his Voyages of Discovery*, 3 vols, (Cambridge, 1955–67)

J. C. Beaglehole, *The Life of Captain James Cook* (London, 1974)

Andrew David, Felipe Fernandez-Armesto, Carlos Novi and Glyndwr Williams, *The Malaspina Expedition 1789 to 1794: Journal of the Voyage by Alejandro Malaspina* (London, 2001–2004)

Greg Dening, *Mr Bligh's Bad Language* (Cambridge, 1992)

Greg Dening, *The Death of William Gooch: A History's Anthropology* (Carlton South, Vic., 1995)

Miriam Estensen, *The Life of Matthew Flinders* (Crows Nest, 2003)

Robin Fisher and Hugh Johnston (eds), *Captain James Cook and His Times* (Vancouver, 1979)

Robin Fisher and Hugh Johnston (eds), *From Maps to Metaphors: The Pacific World of George Vancouver* (Vancouver, 1993)

John Gascoigne, *Captain Cook: Voyager Between Worlds* (London, 2007)

Derek Howse and Beresford Hutchinson, *The Clocks and Watches of Captain James Cook* (AHS, 1969)

Wayne Orchiston, *Nautical Astronomy in New Zealand: The Cook Voyages* (Wellington, 1998)

E. G. R. Taylor, *Navigation in the Days of Captain Cook* (London, 1974)

Glyndwr Williams, *Captain Cook: Explorations and Reassessments* (Woodbridge, 2004)

Chapter 6

William J. Ashworth, 'The calculating eye: Baily, Herschel, Babbage and the business of astronomy', *British Journal for the History of Science*, 27 (1994), 409–41

John Brooks, 'The circular dividing engine: Development in England 1739–1843', *Annals of Science*, 49 (1992), 101–35

Andrew Cook, 'Alexander Dalrymple and John Arnold: Chronometers and the representation of longitude on East India Company charts', *Vistas in Astronomy*, 28 (1985), 189–95

Richard Dunn, 'Scoping Longitude: Optical Designs for Navigation at Sea', in G. Strano et al. (eds), *From Earth-Bound to Satellite: Telescopes, Skills and Networks* (Brill, 2011), 141–54

Humphrey Jennings, *Pandæmonium: The Coming of the Machine as Seen by Contemporary Observers* (London, 1995)

Anita McConnell, *Jesse Ramsden (1735–1800): London's Leading Scientific Instrument Maker* (Aldershot, 2007)

A. D. Morrison-Low, *Making Scientific Instruments in the Industrial Revolution* (Aldershot, 2007)

Simon Schaffer, 'Babbage's Intelligence: Calculating Engines and the Factory System', *Critical Inquiry*, 21 (1994), 203–27

Doron Swade, *Charles Babbage and his Calculating Engines* (London, 1991)

Chapter 7

Philip Arnott, 'Chronometers on East India Company Ships 1800 to 1833', *Antiquarian Horology*, 30 (2007), 481–500

John Cawood, 'The Magnetic Crusade: Science and politics in early Victorian Britain', *Isis*, 70 (1979), 492–518

Archibald Day, *The Admiralty Hydrographic Service 1795–1919* (London, 1967)

Harry W. Dickinson, *Educating the Royal Navy: Eighteenth- and Nineteenth-Century Education for Officers* (Abingdon and New York, 2007)

Fergus Fleming, *Barrow's Boys: A Stirring Story of Daring Fortitude and Outright Lunacy* (London, 2001)

Jordan Goodman, 'The hell-borne traffic: William Owen and the African slave trade', *Geographical*, 79 (2007), 63–65

R. D. Keynes, *The Beagle Record: Selections from the Original Pictorial Records and Written Accounts of the Voyage of HMS Beagle* (Cambridge, 1979)

W. E. May, *How the Chronometer Went to Sea* (AHS, 1976)

D. P. Miller, *The Royal Society of London, 1800–1835: A Study in the Cultural Politics of Scientific Organization*, PhD Dissertation (University of Pennsylvania, 1981)

Michael S. Reidy, *Tides of History: Ocean Science and Her Majesty's Navy* (Chicago, 2009)

G. S. Ritchie, *The Admiralty Chart: British Naval Hydrography in the Nineteenth Century* (London, 1967)

INDEX

Page numbers in bold refer to captions, diagrams and photographs.

Académie des Sciences 36, 50, 60, 120
Adams, John 147
Admiralty 45, 77, 82, 93, 114, 124–5, 133, 137, 184, 185, 188, 192, 197, 217
Admiralty Chart 188
Adventure (ship) 122, 128, 131, **134–5**, 210
Africa
    charting 202–7, **203, 204–5**, 206
Airy, George 175–7
Aleutian Islands 137
Alexander (ship) 197
Alimari, Dorotheo 70, **70**
American Philosophical Society 155
Analytical Engine 175–7
Andrews, Henry 113
Anne (Queen) 56, 73
Anson, George 82, 96, 128–9, 137, 216
Apian, Peter
    Cosmographica (1524) 51
    Introductio Geographica (1533) 51, **52**
Apollodoro, Francesco, *Galileo Galilei* (*c*.1602–07) **48**
Arbuthnot, John 76
Arctic 199, 202
Arnold, John 166, 171–2, 175
    John Arnold and family (Davy, *c*.1783) **164**
    marine timekeepers by **116**, 119, 121–2, 133, 150, 154, 155, 166, **167, 212**
        no. 23 (*c*.1784) **167**
        no. 294 (*c*.1807) **216**
Arnold, John Roger 171
    no. 36 (1778) **165**, 166
artificial horizon **94**, 95, 216
Association (ship) 29
Astrolabe (ship) 151
Atlantic Ocean 19
Australia 19, 211, 217
    charts **24**, 25
    coastal survey 150, **214**
    colonisation 145
    shipwreck 30
Austrian Succession, war of the 82

Babbage, Charles 175–7, **176**
    Specimen of Logarithmic Tables (1831) **177**
    see also Difference Engine, Analytical Engine
Backhuysen, Ludolf **16–17**
backstaff 22, **23**, 31, 93, **94**, 95
Baffin, William 53
Baffin Bay 197
Baily, Francis 175
balance spring, helical 166
Banks, Joseph 129, 145, 150, 166, 171–2, 178, 197
Barbados 66, 99–101, **100–1**, 102, 122, 144
Barracouta (brig) 202
Barrow, John 192, 197–8
Batavia (ship) 30
Bayly, William 131, 133
Beagle (ship) **206–7, 208–9**, 208–17, **210–11**
beam compass 173
Beaufort, Francis 214, 220
Bedlam **74–5**, 76
Berthoud, Ferdinand 119–20, 151, 154, 164
    horloge marine no.8 (1767) **117**
Besson, Jacques **99**
Best, William Philip 96
Bidstrup, Jesper 173–5
Billingsley, Case 70
bimetallic strip 88
Bird, John 116, 173
    marine sextant (*c*.1758) 96, **97**
    transit instrument **110**, 112
Bishop, Robert, lunar-distance form (1768) **123**
Blackheath 42, 73
Bligh, William 145–7, **146–7**, 150, 181
Board of Longitude 124, 126, 147, 157–8, 177–8, 192, 199
    and the *Bounty* 145
    and the construction of affordable timekeepers 108–9, 117, 119–22
    and Cook's voyages 133
    disbanded 185, 197
    and Henry Constantine Jennings 184
    and John Arnold 166, 171
    and John Bird 173
    and Jupiter's satellites 181
    and longitude measurement, training in 124–5, 150
    and magnetic variation 181–4
    miscellaneous schemes presented to 185
    and the *Nautical Almanac* 110, 112, 113
    and Ramsden's dividing engine 173
    reworked (1818) 197
    and Thomas Earnshaw 171–2
    and Thomas Mudge 164–6
    and timekeepers 108–9, 117, 119–22
    see also Commissioners of Longitude
Bombay (Mumbai) 193
Bond, Henry 45, 53
Botany Bay 151
Bougainville, Louis-Antoine, Comte de 150–1
Bounty (ship) 145–7, **146–7**
Bourne, William 12
Boussole (ship) 151
Bradley, James 93, 96, 98
Bradley, John 185
Brazil 129
breadfruit 145, **145**, 150
Brest 188
Bridgetown, Barbados **100**, 101
Brisbane, Thomas 156
Britain 31, 50, 65
    chartered companies 16
    Industrial Revolution 158, 162
    manufacturing expertise 186
    see also England
British Mariner's Guide 108, 113, 129
Brocklesby Park 77
Brouncker, William 45
Bruce, Alexander, Earl of Kincardine **58**, 60
Bruhl, Count 120
Brunel, Marc Isambard **174**, 175
Buchan, David 197
Burden, Monsieur 32
Burke, Edmund 122
Busy Body 99

Campbell, John 96
Camus, Charles-Étienne 119, 120
Cape Blanco 28
Cape Colony 202
Cape of Good Hope 19, 25, 99, 133, 150, 193
Cape Guardafui 202
Cape Horn 30, 82, **83**, 96, 131, 210–11, 216
Cape Newenham 140
Cape Noir 82
Cape Spartel 31
Caribbean 25, 28
Carlyle, Thomas 158
Carteret, Philip 147

247

Cassini, Giovanni 50
Cathcart, Charles 186
celatone 45–50
Centurion (ship) 77, **80–1**, 82, **82**, **83**, 96, 128
Cervantes, Miguel de 36
Chandos, Duke of 73
Charles II 31, 53
Charlton Island 28
chart **14–15**, **18–19**, 19, 21–2, **24**, 25, 28, 45, **46–7**, **83**, 128, **129**, 131, **136–7**, **142**, **149**, 151, 154, **204–5**, **208–9**, **214**, **216–7**, **218–9**, 221
   unreliable 25, 31, 137
charting 21, 137–44, 202–7
Chatham (yacht) 93, 150
Chavasse, William **179**, 181
China 150, 181, 186–7
chinaware 17
Christian, Fletcher 145–7
chronometer 73, 156–7, 162–72, 186–7, 189, 192, 193, 197–8, 206–7, 214, 216, 221
   see also specific makers
Cinque Ports (ship) 30
Clairaut, Alexis 96
Clark, William 155
Clement, Joseph 175
clock 39, 57–63, **58–62**, 66, 70–3, 172
   bimetallic strip 88
   caged roller-bearing **86**, 88
   effects of motion at sea on 77
   friction 77, 88
   grasshopper escapement 77, **78**, 88
   gridiron pendulum 77, 88
   Harrison's **76**, 77–89, **78**, **82**, **85**, **87**, 92
   isochronism 77
   pendulum 57–8, 63, **76**, 77
   remontoire **85–7**, 88
   temperature compensation 77, 88, 166
   Thacker's **72**, 73–6
   see also specific makers
coal 158
coffee-houses 69, **71**, 73
Collins, Greenvile, *Great Britain's Coasting Pilot* (1693) **24**, 25
Colson Nathaniel, *New Seaman's Kalendar* 31
Columbus, Christopher 28
Commission for the Discovery of Mr Harrison's Watch 116
Commissioners of Longitude 39, 42, 45, 53, 65–6, 69–73, 77, 125, 178

and the 1765 Longitude Act 104
and the construction of affordable timekeepers 116–17, 119, 122
and Irwin's marine chair 99, 104, 108, 114
and the Harrisons 82, 88, 92, 101–3, 114, 122
making longitude techniques available 104, 108, 114
and Nevil Maskelyne 101–2
and timekeepers 116–17, 119, 122
and Tobias Mayer 96, 99
see also Board of Longitude
Compagnie Française pour le Commerce des Indes Orientales (French East India Company) 17
compass (beam) 173
compass (magnetic) 22, **22**, **94**
   amplitude compass (1780) **43**, 45
   azimuth compass (*c*.1793) 181–4, **182**, 198
   insulating compass (*c*.1818) 184, **184**
compensation balance 166
Connaissance des Temps 98, 99, 108, 151, 154
Cook, James 119, 122, 128–33, 137–44, 147, 188, 197
   Captain James Cook (Dance, 1775–76) **128**
   'Chart of the Island of Otaheite [Tahiti]' (1769) 129
   Chart of New Zealand (1772) **142**
   Copley Medal 88
   death 144, 145
   first voyage 128–31
   journal on *Resolution* **131**
   second voyage 122, 126, 128, **130**, 131–3, **132–3**, 137, 140–1, **142**, 145, 166
   third voyage 128, **128**, 137–44, **138**, **143**, 145, 150
Couch, John 185, **185**
Council of the Indies 25
Croaker, Henry 185
Croker, John Wilson 197
Crosley, John 150, 171
cross-staff 22, 51, **53**, 93
Cruickshank, George, 'Landing the Treasures ...' **198–9**, 199
currents 30, 31

Daily Courant (newspaper) 73
D'Alembert, Jean 96

Dalrymple, Alexander 154, 166
Dampier, William 25, 30, 82, 129
Dance, Nathaniel
   Captain James Cook (1775–76) **128**
   Thomas Mudge (*c*.1772) **162**
Dandridge, Bartholomew, *John Hadley* (early 1730s) **92**
Darwin, Charles 210
Davy, Robert, *John Arnold and family* (*c*.1783) **164**
dead reckoning 22, 25, 28, 31, 45, **82**, 98, 131, **131**, 147, 156, 157, 178, 192
Defoe, Daniel 69, 162
   Robinson Crusoe (1719) 30
Demainbray, Stephen 121
Deptford (ship) 92
depth sounding **24**, 25, 178
Desaguliers, John 73
Descubierta (ship) 154
D'Evreux de Fleurieu, Charles-Pierre 120
Dickens, Charles, *Dombey and Sons* (1848) 175
Difference Engine 175–7, **176**
dip circle **138**, 202, **202**
Discovery (ship of Cook's expeditions) 128, **140–1**
Discovery (ship of Vancouver's expeditions) 150
disease 202, 207
Ditton, Humphry 39–42, 45, 57–8, 73, 76
dividing engine **171**, 173–5
Dodd, Robert, 'The Mutineers turning Lt Bligh and part of the officers and crew adrift' (1790) **147**
Dollond 154, 187
Dolphin (ship) 129
Dorothea (ship) 197
double reflection
   principle of 93–5, 96
   instrument 93–5, **93**
Drake, Francis 99
Dryden, John, *Annus Mirabilis* (1667) 19, 22
Du-Val, Pierre, *Carte universelle du commerce* (1686) **14–15**
Dunthorne, Richard 110
   Tables Requisite ... **109**, 110–12, 147

Eagle (ship) 29
Earle, Augustus 210
   'Life on the ocean ...' (*c*.1820–37) **194–5**

Earnshaw, Thomas 150, 166–72, 175
    escapement model (1804) **169**, 171
    no. 512 (*c.*1800) **168**
    no. 524 (*c.*1800) **168**
    Thomas Earnshaw (Schee, *c.* 1808) **166**
Earth, as timekeeper 57
Earth's magnetism 42–5
    magnetic inclination 43–5, 73, 202
    magnetic variation 43–5, **46–7**, 73, 181–4
East India Company 16–17, 98–9, 156–7, 166, 184, 186, 192–3
Eckebrecht, Philipp 25
eclipses
    Jupiter's satellites 39, 45–51, 53, 92, 99
    lunar 25–8, 45, 96, 112
    solar 25–8, 45, 73, 112, 129
Edwards, Eliza 114
Edwards, John 113–14
Edwards, Mary 113–14
electric telegraph 221
Elliot, John 95
Elliott, John 126
Elton, John, backstaff with artificial horizon **94**, 95
Emerson, William 125
Emery, Josiah 187
Endeavour (ship) 128, 129–30
England 16, 36
    trade 16–17, **17**, 19
Englefield, Sir Henry 198
English Channel 25, 31
equal altitudes 93
Equator 20–1, 112
escapement
    Earnshaw's **169**, 171
    grasshopper 77, **78**, 88
    spring-detent 166, 171
Euler, Leonhard 96, 103

Faraday, Michael 197
Fernando Po 207
Ferrner, Bengt 99
Ferro (El Hierro), Canary Islands 20
fever 202, 207
Fidler, Peter 155
Firebrand (ship) **28–9**
FitzRoy, Robert 196, **208–9**, 210–16, **211**, **215**
Flamsteed, John 21, 31, 50, 53–7, 70, 73, 77

flax 144
Flinders, Matthew 150, 156, 188, 202
Flinders, Samuel 150
'fluid quadrant' 77
Folkes, Martin 73
Forster, Georg 144
France 16, 36, 109, 126, 150–4
    clock and watchmaking 63, 119–21, 151, 154, 166
    coastline maps 50, **51**
    and the lunar-distance method 51
    monarchy 60
    navigation schools 31
    trade 19
Franklin, Benjamin 25, 89
Franklin, John 197, 199, **200**, 202
French, John 73
French Revolution 145, 173
friction, reduction of 77, 88
Funnell, William 30

Galapagos Islands 25
Galilei, Galileo **48**
    clocks 57
    journal **49**
    observations of Jupiter and its satellites 45–50, **49**
Gellibrand, Henry 28
Gentleman's Magazine 105
George II 73, 96
George III 117, 121–2, 125, 129, 166
George, Prince of Denmark 56, 73
Gilbert, Joseph, 'Part of the Southern Hemisphere...' (*c.*1775) **137**
Gilbert, William, *De Magnete* (1600) 43
Glendon, John 13
Godfrey, Thomas 93–5, **93**
Graham, George 63, 77, 164, 173
    astronomical regulator **110**, 112
grasshopper escapement 77, **78**, 88
gravity, inverse square law of 96
Great Lakes 202
Green, Charles 99–101, 129–31, 141
Greenland 53
Greenwich Hospital for Seamen 31, 197
Greenwich Meridian 20, 188, 221
Greenwich Observations 112
Greenwich time 109, 193, **212**
gridiron pendulum 77, 88
Griper (ship) 199

Hack, William 19, **18–19**
Hadley, George 93
Hadley, Henry 93
Hadley, John **92**, 93–5
    John Hadley (Dandridge, early 1730s) **92**
Hadley quadrant/octant **92**, 95, 98, 124, **139**
Hague, The 60
Halley, Edmond 31, 43, 45, 56
    and double reflection 93
    Edmond Halley (Murray, *c.*1690) **44**
    and John Harrison 77
    and Whiston's rocket scheme 73
    world chart showing magnetic variation (1702) **46–7**
Hamilton, Archibald **94**
Hampstead Heath 42
Hanmer, Thomas 73
Harper's New Monthly Magazine 176
Harrison, James 77
Harrison, John 66, 77–92, 99, 101–3, 104, 125, 162, 164, 166, 173
    John Harrison (King, *c.*1765–66) 89, **89**
    H1 (1735) 77, **78–9**, 82, **82**, 89
    H2 (1737–39) **84–5**, 88, 89
    H3 (1740–59) **86–7**, 88, 88–9, 92
    H4 (1755–59) 89–92, **90–1**, 101, 102–3, 108, 114–17, **114**, 119–20, 121, 162, 164
    H5 (1770) 120, 121–2, **121**
    medallion portrait (Tassie, *c.*1776) **122**
    precision long-case regulator (1726) **76**, 77
    Principles of Mr. Harrison's Timekeeper (1767) **115**, 119, 120–1
Harrison, John (grandson of John) 121
Harrison, John (ship's purser) 129
Harrison, William **91**, 92, 101–2, 121–2
Hawkesworth, John, *An Account of the Voyages Undertaken ... in the Southern Hemisphere* (1773) **145**
health at sea 22
Heath, William, 'March of Intellect ...' (1829) **160–1**
Hecla (ship) 199
Herschel, John 175, 220–1
Hipparchus of Nicaea 20
Hitchens, Malachy 113, 114
Hobbs, William 73
'Horologe' 73

Hodges, William
    Cook's *Resolution* in the Marquesas Islands (1774) **130**
    'View of Maitavie Bay' (1776) **134–5**
    View of Point Venus and Matavai Bay, Tahiti (1773) **132–3**
Hodgson, James 73
Hogarth, William 89
    A Rake's Progress **74–5**, 76
Holland 65
    see also Netherlands
Holmes, Robert 60
Home, Robert, *Jesse Ramsden* (*c.*1791) **170**, 173
Hooke, Robert 45, 53, 60, 93
horizon 22
Hornsby, Thomas 164
House of Commons Committee 39, 43
Houtman, Frederik de 30
Houtman Abrolhos 30
Howe, Lord 99
Howells and Pennington, no. 4 (*c.*1794) **163**
Howells, William 166
Hudson's Bay Company 16, 155
human error 28, 31
Hunter, John 150
Huygens, Christiaan 57, **58**, 60–3, **60**, 77
Hydrographic Office 188, 197
hydrographic surveying *see* surveying

Illustrated London News **192**
India 17–19, 25
Indian Ocean 16, **24**, 30, 202
Indonesia 19
Industrial Revolution 158, 162
Inman, James 150
innovation 158, **161**, 162, 175
International Meridian Conference, Washington D.C. (1884) 221
Investigator (sloop) 150
Irwin, Christopher 66, 99–101
    marine chair 66, 99, 101, 181
Isabella (ship) 197
isochronism 77

Jamaica 92
James II 19
James, Thomas 28
Java 30

Jefferson, Thomas 155
Jefferys, John 89, 117
Jennings, Henry Constantine 184–5
    insulating compass (*c.*1818) 184, **184**
    mercury log glass (*c.*1817) **183**, 184
Jervas, Charles, *Isaac Newton* (1717) **38**
Jervis, John 145
Johnson, Samuel 158
Juan Fernandez archipelago 30
Jupiter's satellites 73, 99–101, 109, 137
    Chavasse's observing platform (1813) **179**, 181
    eclipses 39, 45–51, 53, 92, 99
    marine chair for the viewing of 66, 99, 101, 181
    Parlour's shoulder-mounted apparatus for observing (1824) **180**, 181

Kamchatka 151–4
Kater, Henry 198
Kealakekua Bay, Hawaii 144
Keech, Joseph 113
Kempthorne, Rupert 13
Kendall, Larcum 116
    K1 (1769) **116**, 117–19, 121–2, 133, **139**, 140–1, 145
    K2 (1771) 119, 145–7, **145**
    K3 (1774) 119, 121, 141, **148**, 150, 154, **154–5**
King, John Lawrence, log of the *Owen Glendower* **212–13**
King, James 137–40, 141–4
King, Philip Parker 210, 214
King, Thomas, *John Harrison* (*c.*1765–66) 89, **89**

Lacaille, Nicolas Louis de 99, 108
Laccadive Islands (Lakshadweep) 25
Lalande, Jérôme 119, 120
Lapérouse, Jean-François de Galaup, Comte de 151, **152–3**, 154
latitude 12, 73, 92, 93, 216
    definition 20–1, **20**
    errors in the measurement of 31
    latitude sailing 25, 30, 133
    measurement 20–2, **20–1**, **23**, 25, 42–5, 109, 184
    plotting geographical position by 20, **21**

Le Dieu, Franciscus 61
Le Roy, Pierre 166
    montre marine **118**, 120
lead and line **24**, 25
Lenoir, Étienne 173
Lesseps, Jean-Baptiste Barthélemy de 151–4
Leven (sloop) 202
Lewis, Meriwether 155
lignum vitae 77, **78–9**, 85–7
Lisbon 60
    trial of H1 to 77, 82
Livorno 50
Lochard, Lieutenant **23**
log
    log book 22, **23**, 31, 82, **82**, 157, 212–13
    chip-log (ship-log) 22, 25, 31
    mechanical 178, **178**
Logan, James 93
London Magazine 99
long-distance travel 12, 16–19, 25, 96–101
longitude
    by dead reckoning 22, 25, 28, **82**, 98, 131, **131**, 178
    definition 20–1, **20**
    determining at sea 21, 25–8, 32, 76, 96, 98, 133, 156, **212–13**, 221
    determining on land 21, 25
    by Ditton and Whiston's method 39–42, 45
    early measurement attempts 25–8
    errors in the measurement of 28
    importance of 16–19
    by Jupiter's satellites 39, 45–51, 53, 66, 73, 92, 99–101, **99**, 109, 137, **179–80**, 181
    by magnetic inclination 43–5, 73, 202
    by magnetic variation 43–5, **46–7**, 73, 181–4
    octant, use for determining **92**, 93–5, **94**
    plotting geographical position by 20, **21**
    problem of determining 12, 19, 21, 22, 32–3, 34, 36, 76
    rewards for a solution to the problem of 36–9, 65, 69–70, 73, 76, 77, 82, 99, 101, 102–3, 104, 108, 114, 116, 121, 122, 164, 166, 173, 199
    see also lunar-distance method of longitude measurement; timekeeping method of longitude measurement

Longitude Act 164
  1714 (British) 33–4, 36–9, **37**, 53, 65–6, **68**, 69–73, 76, 82, 96, 102–4, 108, 114, 121–2, 197
  1765 (British) 103, 104, 117, 122
lookout 22, 25
Louis XIV 50
Ludlam, William 116, 117, 166
lunar occultation 96
lunar distance
  form to aid calculation **123**
  French trials 151
  method of longitude measurement 39, 51–7, 66, 92–3, 95–101, 103–4, 117, 126, 150, 156–7, 178, **212–3**
  use on the Cook expeditions 129–31, **131**, 133, 137–44
Lyons, Israel 110
Lyttelton, George 76

Macartney, Lord George 186–7
Madagascar Strait 19
Madeira 92
Madras (now Chennai) 193
magnetic inclination 43–5, 73, 202
magnetic variation 43–5, **46–7**, 73, 181–4
Magnificent (ship) 188, **192**
Malaspina, Alejandro 154–5
Maldives 25
Manila galleons 17
Maori 131
Mapson, John 110
Margaret, Lady Clive 114
Margetts, George, marine chronometer (*c*.1790) **203**
marine timekeepers
  affordable 157, 158, 162–4
  value 145–7
  see also chronometer; clock; watch
  see also specific makers
Mars 28
Marshall, John, navigational workbook 193–6, **196**
Martens, Conrad 210–11
  The *Beagle* in Beagle Channel (*c*.1834) **206–7**
  'Portrait Cove, Beagle Channel' (*c*.1834) **210–11**
Maskelyne, Nevil 98–102, 124
  and the construction of affordable timekeepers 108–9, 116–17, 119
  and Cook's voyages 129, 131–3, **138**
  death 175, 197
  and magnetic variation 181
  and Mr Harrison's Watch 116, 122
  and the *Nautical Almanac* 109–14, **112–13**, 192
  Nevil Maskelyne (Russell, *c*.1776) **98**
  Nevil Maskelyne (van der Puyl, 1785) **111**
  and the observation of Jupiter's satellites 101, 109, 181
  observing suit **110**, 112
  and the *Tables Requisite* **109**, 110–12, 147
  and *The British Mariner's Guide* (1763) 98–9, **98**
  and Thomas Earnshaw 171–2
  and Thomas Mudge 164, 166
  and timekeepers 108–9, 116–17, 119
mass production 172, 175
Massey, Edward 178
  mechanical log **178**
Matavai Bay, Tahiti **132–3**, **134–5**
Matthews, William 116
Maudslay, Henry **174**, 175
Mauritius 150, 193, 202
Mayer, Tobias 66, **94**, 95–9, 101–3, 104
  Germaniae atque in ea Locorum Principaliorum Mappa Critica (1750) **95**, 96
  repeating circle 96, **96**
mechanical log 178, **178**
mechanization 162–77, 187
Medici family 50
mégamètre 151
meridian **14–15**, 16, 20, 214, 221
Merlin (ship) 92
Mexico 19
Michaelis, Johann David 96
Mitchell, John 116
Mombasa 207
Montevideo 214
Moon
  eclipses 25–8, 45, 96, 112
  effect of the Sun on the motion of 56, **56**, 96
  Mayer's map 96
  transits 112
  see also lunar-distance method of longitude measurement
moondial 25
Moore, John Hamilton 147, 196
Moore, Jonas 53
Moore, Joshua 113
Morin, Jean-Baptiste 51
mortars **41**, 73
mortising machine **174**, 175
Mouat, Thomas 166
Mozambique 202, 206
Mudge, Thomas 116, 120, 162, 164–6
  'Blue' 164
  'Green' (1777) **162**, 164
  Thomas Mudge (Dance, *c*.1772) **162**
Mudge, Thomas junior 164, 166
Murray, Thomas, *Edmond Halley* (*c*.1690) **44**
Muscovy Company 16

Napoleonic Wars 188
Narborough, John 28
natural hazards 17
nautical almanac 104, 108
  Nautical Almanac and Astronomical Ephemeris 104, 108–14, **108–9**, 124, 131, **138**, 145, 154–5, 157, 175, 188, 192, 197, **212**, 214
Nautical Magazine 192–3
navies 19, 154
  see also Royal Navy
navigation 12
  by animals, birds and plants 22, 25
  by coastal features 22
  by man-made features 22
  drive to improve navigational knowledge 31
  mistakes 28
  pilots (rutters) 22
  Polynesian 131
  practice of 22–5
  workbook (Marshall) 193–6, **196**
  see also dead reckoning; Jupiter's satellites; latitude; longitude; lunar distance; timekeeper method
Navy Board 150
Netherlands 16, 30
  longitude research 36, 43
  trade 17, **17**, 19
  see also Holland
New Holland (Australia) 150

New South Wales 150
New Spain (Mexico) 17
New Zealand 131, 141, **142**
Newton, Isaac 31, 32, 39
    and double reflection 93
    and Flamsteed 56
    inverse square law of gravity 96
    Isaac Newton (Jervas, 1717) **38**
    and the measurement of longitude 42–3, 45, 56–7, 70–3
    Philosophiae Naturalis Principia Mathematica 56, **56**, 96
    and the three-body problem 56, 96
Niebuhr, Carsten 96–8
Nootka Sound **128**, 140, **140–1**
    Crisis 150
Norfolk Island 145
North, Lord 122
North America 155
north magnetic pole 202
North Pole 20, 145, 197, 199
North Sea 25
North-West Passage 28, 53, 128, 150, 184, 197–9
Norton Sound 140
Nunavut, Canada 28

observing platform (Chavasse, 1813) **179**, 181
occultations, lunar 96
octant 96, 147, **194**, **212**
    development **92**, 93–5, **94**
    Hadley quadrant/octant **92**, 95, 98, 124, **139**
    scales 172–5
    see also Godfrey, Thomas, Hadley, John
oil 92
Oman, Sultan of 207
Oosterwijck, Severyn **58**, 60
Orford (ship) 82
Owen, Richard 189, 206–7
Owen, William Fitzwilliam 202, 206–7
Owen Glendower, log of the **212–13**

Pacific islands 214
Pacific Ocean 16, 19, 28, 30, 82, 119, 128–33, 137, **137**, 150–1, 154
palm oil **205**, 207

parallel rule 211, **211**
Paramore (ship) 45
Paris Observatory 36, 50, **50**, 51
Parlour, Samuel **180**, 181
Parramatta Observatory 214
Parry, Edward 197, 199, **200**
Patagonia 210
Patterson, Robert 155
pendulum 57–8, 63, **76**, 77
Pennington, Robert 166
Pepys, Samuel 22, 31
    Tangier Papers (1683) 19
Philadelphia 25
Philip II of Spain 36
Philip III of Spain 36
Philip, Arthur 145, 151
Philippines 17, 19
Philosophical Transactions (journal) 25, 129
Phipps, Constantine 145
Pitcairn Island 147
Place, Francis, *The Octagon Room at the Royal Observatory* (*c*.1676) **57**
plants 22
Point Venus, Tahiti **132–3**, **134–5**
polar exploration 197–202
Pole Star 21, 22
Pond, John 113, 114, 192
Porpoise (ship) 150
Port Desire 28
Port Essington 217
Port Jackson 150
Port Royal, Jamaica 92
Portsmouth dockyard **174**, 175
Portsmouth harbour 192
Portugal 16, 17–19, 31
    trade 19
Post Man 73
Price's candles **205**, 207
Prime Meridian 221
    see also Greenwich Meridian
Prince Henry (ship) 98
Princess Louisa (ship) 101
Principles of Mr. Harrison's Timekeeper (Harrison, 1767) **115**, 119, 120–1
privateers 17, 30, 128
    British 19
    French 25
Proctor, John 82, **82**
Providence (ship) 150
Ptolemy 20

quadrant
    astronomical **139**
    Hadley **92**, 95, 98, 124, **139**
    see also octant

Racehorse (ship) 145
Ra'iatea 129
Ramsden, Jesse 147, 154, 187
    dividing engines of **171**, 173, 175
    Jesse Ramsden (Home, *c*.1791) **170**, 173
    sextant **143**
Regiomontaus, *Ephemerides* 25
remontoire **85–7**, 88
Resolution (ship) 122, 126, 128, **130**, 131, **132–3**, **134–5**, **137**, **140–1**
    Cook's journal **131**
Rio de Janeiro 214
Robbins, Reuben 113
Robertson, John 196
Robinson, John 122
rockets 73, 76, 206
Ross, James Clark 199, 202
    Commander James Clark Ross (Wildman, 1834) **202**
Ross, Captain John 184, 197, 199, **200**
Rowe, Jacob 77
Royal Africa Company 16
Royal Commission 45
Royal Danish Expedition to Arabia 98
Royal Institution 197
Royal Mathematical School, Christ's Hospital 31, 39, 193
Royal Naval Academy 193, 196
Royal Naval College, Portsmouth 192, 196–7
Royal Navy 95, 104, 129, 157, 178, 207, 217, 220
    and the azimuth compass 184
    and the *Beagle* voyages 210, 214
    chronometer stock 192
    First Fleet 145, 151
    and Irwin's marine chair 99
    and magnetic variation 45
    Masters of Navy ships 124–5
    Scientific Branch 197
    shipwrecks 30–1
    and the slave trade 202
    special expeditions 126
    training 31, 150, 188, 193–7
    see also Admiralty

Royal Observatory, Cape of Good Hope 197, 214
Royal Observatory, Greenwich 36, 50–1, 53, 56, **56**, 70, 73, 99, **106**, 188, 221
   and John Arnold's timepieces 166
   chronometer testing 192
   lunar observations 112
   and the *Nautical Almanac* 108
   Octagon Room 57, **57**
   and the production of affordable timekeepers 108–9, 116–17, 119
   and Thomas Earnshaw's timepieces 171
   and Thomas Mudge's timepieces 164
   time ball 192–3, **192**
   under the jurisdiction of the Board of Longitude 197
Royal Society 25, 34, 36, 39, 45, 50, 53, 56, 60, 63, 65, 70, 73, 120
   and the Astronomer Royal 112
   and Cook's voyages 129, 133
   and the Difference Engine 175
   and John Hadley 93–5
   and John Harrison's clocks 77, 88
Rupert's Land 155
Russell, John, *Nevil Maskelyne* (*c*.1776) **98**
Russia 126

St Agnes lighthouse **24**, 25
St George (ship) 30
St Helena 98, 133, 193
St Julian, harbour of 28
St Pierre, Le Sieur de 51, 53
Samgoonoodha harbour 137
Sandwich Sound 140
Saron, Jean-Baptiste Gaspard Bochart de 173
scales, divided 172–5
Scarborough Castle 22
Schee, Martin Archer, *Thomas Earnshaw* (*c*.1808) **166**
Schetky, John Christian, 'Loss of the *Magnificent*, 25 March 1804' (1839) **190–1**
science 34, 220–1
Scilly, Isles of
   chart of **24**, 25
   shipwreck **28–9**, 30–1, 42
scoring machine, **174**, 175
Scriblerus, Martinus (satirical character) 76

sea otter 144, **144**
Selkirk, Alexander 30
Seller, John, *Practical Navigation* (1672) **24**, 31
Senhouse, Joseph 181
Seven Years War 99, 129, 150
sextant 133, **139**, 141, 147, 150–1, 155, 188, **194**, **213**, 214, 216
   by John Bird (*c*.1758) 96, **97**
   by Nathaniel Worthington (*c*.1840) **172**
   by Ramsden (*c*.1772) **143**
   by Worthington & Allen (*c*.1831) **215**
   of John Lort Stokes 211
   scales 172–5
Ship Cove, Nootka Sound **128**, 140, **140**
shipwreck 28–31, **28–9**, 32–3, 42, 82, 188
Shooter's Hill **41**, 73
Shovell, Cloudesley **28–9**, 30–1, 32–3, 42
Sirius (ship) 145
Sisson, Jeremiah 99
Sisson, Jonathan **94**
slave trade 145, 202, 207
Sloane, Hans 73
Smith, Caleb 76
Snellen, William, marine timekeeper **119**, 121
Society for the Encouragement of Arts, Manufactures and Commerce 158, 177–8
Solar System, scale of 98
South America 28, **83**, 154
South Pole 20
South Sea Bubble 69
South Seas 145
Southern Continent 128, 137
Southern Ocean **137**
Spain 16, 19, 25, 31, 36, 43, 65, 126, 128, 154–5
   Manila galleons 17
   navy 154
   and the Nootka Sound Crisis 150
   trade 19
Spanish East Indies (Philippines) 17
Spanish Succession, war of the 39–42
speed of ship, measurement 22, 178, 184
   see also log
spices 17
Sprat, Thomas, *History of the Royal-Society of London* (1667) **59**, 60

spring-detent escapement 166, 171
Squire, Jane 73
stars 53
   clock stars 112
   observations of 112
States of Holland 36
steam power 158, 162, 221
Stokes, John Lort 211–14, **214–15**, 217
Stokes, Pringle 210
storms 17, 19, 28, 30, 82
Strait of Gibraltar 31
Strait Le Maire 131
Streatfeild, Thomas, deck scene (1820) **193**
Stukeley, William 77
Sully, Henry **62**, 63, 77
Sun 21, 22, 45, 93
   eclipses 25–8, 45, 73, 112, 129
   effect on the motion of the Moon 56, **56**, 96
   transit of Venus across 98, 128, 129
   transits of the 112
surveying 50, 137–44, 150, 154–5, 192, 202–7, 211–21, 223
Swallow (ship) 129, 147
Swift, Jonathan 66, 76

Tables Requisite **109**, 110–12, 147, 155
Tahiti 128–9, **129**, 131, **132–3**, **134–5**, 151
   breadfruit **145**, 145, **146–7**
Tartar (ship) 101
Tassie, James, medallion portrait of John Harrison (*c*.1776) **122**
tea 17
telescope 45–50, **50**, 73, 137, **138**, **194**
temperature compensation 77, 88, 166
terrella **42**, 43
Thacker, Jeremy, *The Longitudes Examin'd* (1714) **72**, 73–6
theodolite **215**, 216
Thomson, John, *New General Atlas* (*c*.1830) **200**
three-body problem 56, 96, 98
Ticknor, George 175
tides 31
Tierra del Fuego 30, 82, **208–9**, 210
timber 144
time ball 192–3, **192**
timekeeper *see* chronometer; clock; watch

timekeeper method of longitude
  measurement 39, 57–63, **58–62**, 70–3,
  93, 104, 114–25, **114–21**, 126, 156–7, 178,
  192–3
  affordable timekeepers 108–9, 116–17,
    119–22, 157–8, 162–72
  French trials 151
  John Harrison on **76**, 77–89, **78**, **82**,
    85, **87**, 92, 102–3
  method **212–13**
  Nautical Almanac for 108
  Spanish trials 154–5
  testing on Cook's voyages **131**, 133,
    137–44, 145
  Thacker on **72**, 73–6
  and the value of marine timekeepers
    145–7
  Vancouver's testing of 150
Timor 147
Tompion, Thomas 57
Tories 42
trade 12, 16–19, **17**, 28, 144, 186, 188, 221
trading companies 16–17, 19
  see also specific companies
travel, long-distance 12, 16–19, 25,
  96–101
Trent (ship) 197
triangulation 216
Trueman (sloop) 93
Tupaia 129
Tyburn 158

United States of America 155

Van de Velde, Willem, the Younger **26–7**
Van der Puyl, Louis François Gérard,
  Nevil Maskelyne (1785) **111**

Van Keulen, Johannes, *The Great and
  Newly Enlarged Sea Atlas or Waterworld*
  (1682) 24
Vancouver, George 150, 188
  'A Chart showing part of the Coast of
    N.W. America' (1798) **149**
Vancouver Island **128**
Vanikoro reefs, New Caledonia 151
Venice 65
Venus, transit of 98, 128, 129
Vereenigde Oost-Indische Compagnie
  (VOC) (Dutch East India Company) 17,
  30, 36, 45, 60
Véron, Pierre-Antoine 150–1
Victoria River, Australia **214**

Waddington, Robert 98, 99
Wales, William 110, 131, 133, 157
Walker, Ralph 181–4, **181**
  azimuth compass (*c.*1793) 181–4, **182**
Wallis, Samuel 129
Ward, William 76
warfare 25
watch 39, 57–63, **58–62**, 66, 89, 172
  affordable 108–9
  and Cook's voyages 133
  by John Arnold **116**, 119, 121–2, 133, 150
  by William Snellen **119**, 121
  see also specific makers
water power 158
Wauchope, Robert 192–3
weather 17, 19, 28, 30, 31, 42, 82, 216
Webber, John
  '*Resolution* and *Discovery* in Ship Cove,
    Nootka Sound' (1778) **140–1**
  'Sea Otter' **144**
  'Various articles at Nootka Sound'
    (1776–80) **128**

Werner, Johann 51
West Indies 99, 117, 144, 145
Westall, William, 'Wreck Reef Bank'
  (1803) **151**
Weston, Thomas 31
whales 144, 198
Wheldon, John 66
Whigs 39, 42, 122
Whiston, William 39–43, **40**, 45, 57–8,
  73, 77
  A New Method for Discovering Longitude
    (1714) 39–42
  rocket scheme 73, 76
  The Longitude Discovered (1738) **41**
Whitby 22
Wickham, John Clements 211
Wildman, John R., *Commander James Clark
  Ross* (1834) **202**
Williams, Zachariah 73
Windward Islands 25
Witchell, George 110
Wollaston, William Hyde 198
Wood, John 28
Worthington, Nathaniel
  parallel rule **211**
  sextant **172**
Wreck Reef 150
  'Wreck Reef Bank' (Westall, 1803) **151**
Wren, Christopher 53
Wright, Robert, 'Viaticum Nautarum (The
  Sailor's Vade Mecum)' (1726) **69**

Young, Thomas 192, 197

Zacuto, *Almanach Perpetuum* 25
Zumbach de Koesfelt, Conrad **61**, 63
Zumbach de Koesfelt, Lothar **61**, 63